Research Notes in Mathematics

W9-AGW-598

Submission of proposals for consideration
Suggestions for publication, in the form of outlines and representative
samples, are invited by the editorial board for assessment. Intending
authors should contact either the main editor or another member of the
editorial board, citing the relevant AMS subject classifications. Refereeing
is by members of the board and other mathematical authorities in the
topic concerned, located throughout the world.

Preparation of accepted manuscripts
On acceptance of a proposal, the publisher will supply full instructions
for the preparation of manuscripts in a form suitable for direct photo-
lithographic reproduction. Specially printed grid sheets are provided
and a contribution is offered by the publisher towards the cost of typing.

Illustrations should be prepared by the authors, ready for direct
reproduction without further improvement. The use of hand-drawn
symbols should be avoided wherever possible, in order to maintain
maximum clarity of the text.

The publisher will be pleased to give any guidance necessary during the
preparation of a typescript, and will be happy to answer any queries.

Important note
In order to avoid later retyping, intending authors are strongly urged
not to begin final preparation of a typescript before receiving the
publisher's guidelines and special paper. In this way it is hoped to
preserve the uniform appearance of the series.

Titles in this series

Nondiscrete induction
and iterative processes

F-A Potra & V Ptak

University of Iowa/Mathematics Institute CSAV, Prague

Nondiscrete induction and iterative processes

Pitman Advanced Publishing Program

BOSTON · LONDON · MELBOURNE

PITMAN PUBLISHING LIMITED
128 Long Acre, London WC2E 9AN

PITMAN PUBLISHING INC
1020 Plain Street, Marshfield, Massachusetts 02050

Associated Companies
Pitman Publishing Pty Ltd, Melbourne
Pitman Publishing New Zealand Ltd, Wellington
Copp Clark Pitman, Toronto

© F-A Potra and V Ptak 1984

First published 1984

AMS Subject Classifications: (main) 41A25, 41A60
(subsidiary) 47D99, 65B99, 65D99

Library of Congress Cataloging in Publication Data
Potra, F.-A.
  Nondiscrete induction and iterative processes.

  Bibliography: p.
  1. Iterative methods (mathematics)   2. Induction
(mathematics)   I. Ptak, V.   II. Title.
QA297.8.P68   1984   519.4   84-7641
ISBN 0-273-08627-8

British Library Cataloguing in Publication Data

Potra, F.-A.
  Nondiscrete induction and iterative
  processes.—(Research notes in mathematics; 103)
  1. Iterative methods (Mathematics)
  I. Title   II. Ptak, V.   III. Series
  512.9'4   QA297.8

  ISBN 0-273-08627-8

Reproduced and printed by photolithography
in Great Britain by Biddles Ltd, Guildford

Contents

Preface

Sixteen years have elapsed since the first paper of V. Pták appeared which
inaugurated the method and it seems that the results obtained thus far justify
an attempt to present a more systematic account of the theory. On the following pages the authors present a survey of more than twenty original papers
devoted to nondiscrete mathematical induction and sketch the present state of
the method. This method is nothing more than an attempt to set up a general
theory of iterative constructions in analysis.

The first paper [43] of the series - although its purpose was different -
contained a simple observation from which the theory grew up: it was the
observation that the classical closed graph theorem possesses a simple
quantitative refinement which, in its turn, may easily be formulated even for
nonlinear mappings.

The central idea of the application of the method to the study of iterative
processes of numerical analysis consists in replacing the classical way of
measuring convergence by a more refined one, which makes it possible to give
error estimates that are sharp not only asymptotically but throughout the
whole process.

A rate of convergence is defined as a function, not as a number.

Since a function - as opposed to just one number - may contain more information and thus give a closer estimate for each step of the iteration, it is
plausible that it may also yield a sharp estimate for the final result.

To give the reader a flavour of the difference in this approach, let us
mention the example of a rate of convergence which can be used to describe
the Newton process. The function

$$w(r) = \frac{1}{2} \frac{r^2}{(r^2 + a^2)^{1/2}}$$

where a is a non-negative constant depending on the problem considered - has
the following properties: for very small r it is

$$\sim \frac{r^2}{2a}$$

for large r it is

$$\sim \frac{1}{2} r .$$

The parameter r is to be considered as a measure of the approximation so that r would be small for sufficiently distant steps of the process.

In this manner the function describes the quadratic convergence of the Newton process when r is small - this corresponds, roughly speaking, to the situation when x_n is already sufficiently close to the solution x - as well as the linear convergence in the initial stages of the process.

In the early stages of the theory, V. Pták tried to indicate some directions of work where the application of the method of nondiscrete mathematical induction could lead to some progress. One of the papers, entitled "What should be a rate of convergence?" attempted to justify the departure from tradition in measuring convergence [52].

The Gatlinburg lecture [49] was the first attempt to explain the main ideas by means of an iterative process used for the solution of an eigenvalue problem; the emphasis is on the didactic side of the matter - the paper was intended to be presented orally at the sixth Gatlinburg conference in 1974.

In the paper [50] the rate of convergence of the Newton method was computed for the first time - the results show that even in this classical subject new insight may be gained if the new point of view is adopted.

Since the interest of V. Pták was formerly mainly centered around functional analysis he was only imperfectly acquainted with the vast literature concerning Newton's method and its various modifications. After the first paper appeared a number of colleagues kindly called his attention to a number of important contributions in the field.

Thanks are due to J. Albrecht (Clausthal) to whom we owe a reference to the work of J. Kornstaedt and to L. Collatz (Hamburg) for a number of critical comments and some useful references.

The work would probably never have proceeded further if it were not for the kind interest of two friends and colleagues M. Práger and J. Taufer whose encouragement is gratefully acknowledged.

Since 1979 a fruitful co-operation between V. Pták and F.A. Potra produced new contributions to the theory and finally resulted in the decision to write the present synthesis.

The method to be discussed in this monograph could be described as a new way of viewing iterative processes and the authors present here a collection of examples to illustrate what may be gained if this new point of view is adopted. The selection of the examples is intentionally broad to emphasize the scope of the method as well as the inter-relations between seemingly distant parts of mathematics and, last but not least, the fact that the results may be deduced from an extremely simple general principle.

Let us list some of the advantages of the method.

1. For a number of iterative processes of numerical analysis it has given definitive results - it should be stressed that for the processes to be treated in this monograph the theorems obtained by the method are sharp in the sense that neither the conditions for convergence nor the estimates obtained can be improved in the class of problems considered. The estimates obtained are sharp not only asymptotically but throughout the whole length of the process, even in the initial stages of the iterations.

2. For a number of iterative constructions in analysis the method gives convergence proofs of unusual clarity and simplicity. In some cases improvements of the result itself were obtained; even in those cases where the method only yields a new proof there seems to be a non-negligible gain in elegance and deeper insight.

The reader who is interested exclusively in the results concerning the processes of numerical mathematics might limit himself to the corresponding sections (3-6).

The first two sections are not indispensable for the understanding of this material but the reading might be considerably easier for those who are prepared by having acquainted themselves with a careful description of the first two model examples presented in Section 2 which helps to understand the motivation of the methods of proof. A similar comment applies to the third example which explains the connection with other parts of mathematics and describes the position of the results in the whole framework of mathematics. The reader who will not shun the trouble of reading the whole of Section 2 will be able to gain a deeper understanding of the common principle underlying a number of seemingly unrelated results which appear now as particular cases of a quantitative refinement of the classical closed graph theorem in functional analysis.

The authors are deeply grateful to M. Neumann (Essen) and to J. Webb

(Glasgow) for their most painstaking perusal of the manuscript and correction of a number of mistakes.

February 1984

F-A P
V P

Note to the reader

1. The first six sections should be read consecutively. The second one (and possibly also the first one) may be skipped.

2. After reading Section 1 each of the Sections 7-11 can be read independently.

3. For the convenience of the reader three appendices have been added: Appendix A contains two diagrams illustrating how the successive iterates of a multidimensional rate of convergence are generated; Appendix B collects some results on some notions of nonlinear analysis used in Sections 4 and 5; finally in Appendix C we describe the methodology used for computing some partial sums appearing in Sections 2, 4 and 7.

4. The notions of local and semilocal analysis are used in the sense of Ortega and Rheinboldt [25] while the notion of consistent approximation of the derivative is used in a more restrictive sense than there (see Appendix B).

5. The definitions, propositions, theorems etc. are indexed by two numbers, the first of which indicates the number of the corresponding section. The formulae are indexed only by one number in brackets. Within the same section the formulae are referred to according to this number. If the formula we want to refer to belongs to another section then we denote it by its number preceded by the number of the respective section.

Thus (3.14) means formula (14) from Section 3.

1 Introduction

Motivation

In existence proofs in mathematical analysis and in numerical analysis we often devise iterative procedures in order to construct an element which lies in a certain set or satisfies a given relation. At each stage of the iterative process we are dealing with elements which satisfy the desired relation only approximately, the degree of approximation becoming better at each step.

Let us describe in a few words an abstract model for such situations. A point x^* is to be constructed which belongs to a given set M. We start by replacing the given set M by a family $M(r)$ of sets depending on a small positive parameter r; the inclusion $z \in M(r)$ means - roughly speaking - that z satisfies the relation defining M only approximately, the approximation being measured by the number r. Given a point $x \in M(r)$, we are interested in constructing a better approximation $x \in M(r')$. In what follows we intend to show that, under suitable hypotheses concerning the relations between the sets $M(\cdot)$, a simple theorem may be proved which gives the construction of a sequence converging to a point $x^* \in M$.

The construction is done inductively; given a point $x \in M(r)$, the induction step consists in constructing a better approximation $x' \in M(r')$. Two factors are of importance here: the distance between x' and x and the improvement of the approximation, the ratio r'/r.

Repeating this process we obtain a certain sequence $x_n \in M(r_n)$; the assumptions of the theorem must guarantee that x_n will be a Cauchy sequence and that $r_n \to 0$. Hence it is natural that the assumptions have the form of a relation between $d(x,x')$ and r'/r.

The theorem, the so called induction theorem, is closely related to the closed graph theorem in functional analysis; it could be described as a quantitative strengthening of the classical closed graph theorem. Indeed, the closed graph theorem can be viewed, in a certain sense, as a limit case of the induction theorem for an infinitely fast rate of convergence (see

Theorem 1.15.) The proof of the induction theorem is little more than an exercise; moreover, it is similar to the proof of the closed graph theorem; the interest of the result lies exclusively in its formulation which makes it possible to unify a number of theorems in one simple abstract result.

Rates of convergence or small functions

In the sequel we shall state several variants of the above mentioned result. First we shall investigate the notion of rate of convergence which plays an important role in our approach. In what follows T will always denote either the set of all positive numbers or a half open interval of the form $T = \{t; 0 < t \leq t_0\}$ for some positive t_0.

1.1 Definition *A function w:T → T will be called a rate of convergence (or a small function) on T if the series*

$$t + w(t) + w^{(2)}(t) + \dots \tag{1}$$

is convergent for all t ∈ T. □

We have used the abbreviation $w^{(n)}$ for the n-th iterate of the function w, so that $w^{(2)}(t) = w(w(t))$ and so on. In what follows it will also be convenient to use the notation $w^{(0)}(t) = t$, $w^{(1)}(t) = w(t)$. The sum of the above series will be denoted by s(t). The function s will be called the estimate function corresponding to w. This function obviously satisfies the following functional equation:

$$s(t) - t = s(w(t)); \tag{2}$$

one of the consequences of this fact is the possibility of recovering w if s is given. Indeed we have:

$$w(t) = s^{-1}(s(t) - t)$$

(with the exception of pathological cases).

It also turns out that the functional equation (2) characterizes in some sense the rates of convergence. More precisely we have:

1.2 Proposition *Let w and h be two functions defined on T such that w maps T into itself and h is non-negative. Suppose further that w and h satisfy*

2

the relation

$$h(t) = t + h(w(t)) \tag{3}$$

for all $t \in T$.

Then:

1. w *is a rate of convergence on* T;
2. *if the limit* $h(0) = \lim\limits_{t \downarrow 0} h(t)$ *exists*

then $s(t) = \sum\limits_{0}^{\infty} w^{(n)}(t) = h(t) - h(0)$.

<u>Proof</u> For each $t \in T$ and each natural number n we have

$$t + w(t) + \ldots + w^{(n)}(t) = h(t) - h(w^{(n+1)}(t)) \leq h(t)$$

so that the series $s(t) = t + w(t) + w^2(t) + \ldots$ is convergent. $\square$

In the following section the reader will meet different rates of convergence. Nevertheless it will be good to give some examples now.

1.3 <u>Example</u> Let us consider a number c such that $0 < c < 1$.

Then the function $w(t) = ct$ is a rate of convergence on the interval $\{t; t > 0\}$ and the corresponding estimate function is $s(t) = t/(1 - c)$.

1.4 <u>Example</u> Let a be a non-negative number. Then the function $w(t) = 2^{-1} t^2 (a^2 + t^2)^{-1/2}$ is a rate of convergence on the interval $\{t; t > 0\}$ and the corresponding estimate function is $s(t) = t - a + (a^2 + t^2)^{1/2}$.

The proof of the fact that the function w given in Example 1.3 is a rate of convergence is very simple because the series (1) reduces in this case to a geometric series.

It is not so easy to compute the sum of the series (1) for the function w given in Example 2. The reader can, however, check that the functions w and s given there satisfy (2) and then use Proposition 1.2.

The motivation of introducing the above rates of convergence will be given later on when they will be applied respectively to the study of the simple iteration process (Banach's fixed point theorem) and to the study of Newton's process.

The reader will have observed that we define the rate of convergence - in

contrast to the established usage - as a function, not a number. A justi-
fication of this will be given in Section 6.

Now let us return to our original problem.

The Induction Theorem

Given a metric space (E,d) with distance function d, a point $x \in E$ and a
positive number r, we denote by U(x,r) the open spherical neighbourhood of x
with radius r, $U(x,r) = \{y \in E; d(y,x) < r\}$. Similarly, if $D \subset E$, we denote
by U(D,r) the set of all $y \in E$ for which $d(y,D) < r$. We shall also consider
the closed neighbourhoods $\bar{U}(x,r) = \{y \in E; d(y,x) \le r\}$, $\bar{U}(D,r) = \{y \in E;
d(y,D) \le r\}$. The closure of a set S will be denoted by S^-. If we are given,
for each sufficiently small positive r, a set $A(r) \subset E$, we define the limit
A(0) of the family A(·) as follows

$$A(0) = \bigcap_{s>0} (\bigcup_{r \le s} A(r))^-. \tag{4}$$

It is easy to see that $x \in A(0)$ if and only if there exist a sequence of
positive numbers (r_n) with $\lim_{n \to \infty} r_n = 0$ and a sequence of points (x_n) such that
$x_n \in A(r_n)$ for all n and $\lim_{n \to \infty} x_n = x$.

In most cases the simple notation A(0) will be sufficient; whenever this
notation might lead to a misunderstanding we shall denote the above set by

$$\lim A(\cdot) \text{ or } \lim_{r \to 0} A(r)$$

The following two observations will help to clarify the intuitive meaning of
this definition and will also be useful in the sequel. For simplicity, we
shall frequently adopt the following convention: if we are given, for each
sufficiently small positive r, a set $A(r) \subset E$, we shall use, for the whole of
this family, the term "approximate set".

1.5 <u>Remark</u>   *Suppose that the approximate set W(·) is monotone, i.e.*
$W(r_1) \subset W(r_2)$ *if* $r_1 < r_2$. *Then*

$$W(0) \supset (\bigcap W(s))^-.$$

<u>Proof</u> For each s > 0 the union $\bigcup_{r \le s} W(r)$ equals W(s). Hence

4

$$W(0) = \bigcap_{s>0} W(s)^- \supset (\bigcap W(s))^-.$$

1.6 <u>Remark</u> *Let M be a given set and define an approximate set M(·) as follows*

$$M(r) = \{x \in E: d(x,M) < r\} \tag{5}$$

Then $\quad M(0) = M^-.$

<u>Proof</u> Since $M(r)$ is monotone, we have $M(0) \supset (\bigcap M(r))^- \supset M^-$. On the other hand, if $x \in M(0)$ then there exists a sequence of positive numbers $r_n \downarrow 0$ and a sequence $x_n \in M(r_n)$ such that $d(x_n,x) \to 0$. It follows that there exists a sequence $m_n \in M$ with $d(x_n,m_n) < r_n$. Hence

$$d(m_n,x) \leq d(m_n,x_n) + d(x_n,x) \to 0$$

so that $x \in M^-$.

1.7 <u>Proposition</u> (The Induction Theorem). *Let (E,d) be a complete metric space. Let T denote, as before, either the set of all positive numbers or a half open interval of the form $\{t; 0 < t \leq t_0\}$, and let w be a rate of convergence on T. For each $t \in T$ let $Z(t)$ be a subset of E; denote by $Z(0)$ the limit of the family $Z(\cdot)$. Suppose that*

$$Z(t) \subset U(Z(w(t)),t)$$

for each $t \in T$. Then

$$Z(t) \subset U(Z(0), s(t))$$

for each $t \in T$.

<u>Proof</u> Suppose that $x \in Z(t)$. Since $Z(t) \subset U(Z(w(t)),t)$ there exists an $x_1 \in U(x,t) \cap Z(w(t))$.

Now $x_1 \in Z(w(t)) \subset U(Z(w^{(2)}(t)), w(t))$, so that there exists an $x_2 \in U(x_1,w(t)) \cap Z(w^{(2)}(t))$.

Continuing this process we obtain a sequence x_n such that $x_{n+1} \in U(x_n, w^{(n)}(t)) \cap Z(w^{(n+1)}(t))$; it follows that $d(x_n, x_{n+1}) < w^{(n)}(t)$, so that x_n is a Cauchy sequence. Since (E,d) is complete, this sequence converges to a

5

limit x^*. Since $x_n \in Z(w^{(n)}(t))$ and $w^{(n)}(t) \to 0$, we have $x^* \in Z(0)$. Further-more $d(x,x^*) \leq d(x,x_1) + d(x_1,x_2) + \ldots < t + w(t) + w^{(2)}(t) + \ldots = s(t)$, so that $x \in U(x^*,s(t)) \subset U(Z(0), s(t))$. The proof is complete. □

The following version of the induction theorem can also be useful.

1.8 Proposition (i) *Let v be a function which maps the interval T into it-self and such that $v^{(n)}(t)$ tends to zero for all $t \in T$. Let g be a positive increasing function on T such that*

$$s_g(t) = \sum_0^\infty g \circ v^{(n)}(t) < \infty \qquad (6)$$

for each $t \in T$. Then $g \circ v \circ g^{-1}$ is a rate of convergence on $g(T)$.

(ii) *Given a family $M(t)$, $t \in T$ of subsets of a complete metric space E such that*

$$M(t) \subset U(M \circ v(t), g(t))$$

for each $t \in T$, then

$$M(t) \subset U(M(0), s_g(t))$$

for each $t \in T$.

Proof Set $Z(t) = M(g^{-1}(t))$ and apply Proposition 1.7 to the family $Z(\cdot)$ and the rate of convergence $w = g \circ v \circ g^{-1}$. □

We can give the induction theorem another formulation which is simpler formally. For this let us introduce a "metric" in exp E as follows: if A,B are two subsets of E, we set

$$d(A,B) = \inf \{r ; A \subset U(B,r)\};$$

this distance, of course, is not symmetric and may be infinite. Using this concept, we may reformulate the induction theorem in the following way:

"If $d(Z(t), Z(w(t))) \leq t$ for all $t \in T$

then $d(Z(t), Z(0)) \leq s(t)$ for all $t \in T$"

This might be easier to remember although less convenient to apply directly in this form. For the proof it is sufficient to observe that the induction theorem remains true if we replace the open neighbourhoods by closed neighbourhoods. This remark will be repeatedly used in what follows.

The application of the induction theorem does not require formally an explicit algorithm for constructing an $x' \in Z(w(r))$ given an $x \in Z(r)$. There are, however, important cases where such algorithms are naturally available. In this case the induction theorem can be given an equivalent formulation which provides a suitable framework for the study of iterative algorithms.

To give a precise content to the notion of meaningfulness of an iterative algorithm it will be necessary to explain carefully some terminology. If X and Y are two sets we define a mapping R from X into Y as a subset $R \subset X \times Y$ such that, for each $x \in X$ there is at most one $y \in Y$ such that $[x,y] \in R$. The projection of R onto the first set is called the domain of definition of R and will be denoted by $D(R)$. If $X = Y$ we may form iterates of R: thus, for instance, R^2 is the set of all pairs x,y such that $[x,z] \in R$ and $[z,y] \in R$ for a suitable z. Clearly $D(R^2) \subset D(R)$ and $D(R^2)$ may be empty even if $D(R)$ is not. Now let G be a mapping from a set E into itself and let $x_0 \in E$ be given. We shall say that the iterative algorithm (G, x_0) defined by

$$x_{n+1} = Gx_n \quad n = 0,1,2,\ldots \tag{7}$$

is meaningful if x_0 belongs to the intersection of all $D(G^n)$, $n = 1,2,\ldots$.

1.9 Proposition *Let G be as above and let $x_0 \in D = D(G)$. If we can attach to the pair (G, x_0) a rate of convergence w on an interval T and a family of subsets $Z(t) \subset D$, $t \in T$ such that:*

(i)   *$x_0 \in Z(r_0)$ for some $r_0 \in T$,*

(ii) *$Gx \in Z \circ w(t)$ and $d(Gx,x) \leq t$*

for all $t \in T$ and $x \in Z(t)$, then

1. *The iterative algorithm (G, x_0) is meaningful and the sequence (x_n) produced by it converges to a point $x^* \in Z(0) \subset \bar{D}$.*

2. *The following relations are satisfied:*

$$x_n \in Z \circ w^{(n)}(r_0), \tag{8}$$

$$d(x_n, x_{n+1}) \leq w^{(n)}(r_0),$$ (9)

$$d(x_n, x^*) \leq s \circ w^{(n)}(r_0);$$ (10)

3. *Suppose that, for some* $n \in N$ *and some* $d_n \in T$ *we have*

$$x_{n-1} \in Z(d_n);$$ (11)

then

$$d(x_n, x^*) \leq s(d_n) - d_n.$$ (12)

<u>Proof</u> The assumption (ii) may be rewritten in the form of two inclusions

$$G\ Z(t) \subset Z \circ w(t)$$

$$Z(t) \subset \bar{U}(GZ(t), t)$$

for all $t \in T$; we have thus

$$Z(t) \subset \bar{U}(Z \circ w(t), t)$$

for all $t \in T$ and the Proposition reduces to a particular case of the Induction Theorem. □

However let us remark that point 3. of Proposition 1.9 may be obtained as a consequence of point 2. in the following sense; having proved (10) it follows that in particular (for $n = 1$) we have proved the following impli-cation:

"if $x_0 \in Z(r_0)$ then $d(G\ x_0, x^*) \leq s \circ w(r_0)$".

Taking in the above implication x_{n-1} and d_n in place of x_0 and r_0 we obtain point 3.

Let us also remark that x^* is a fixed point of G if $x^* \in D$ and if G is continuous at x^*.

Point 3. of Proposition 1.9 will be generally used with the choice $d_n = d(x_n, x_{n-1})$. In this case we shall obtain estimates of the form

$$d(x_n, x^*) \leq s_1(d(x_n, x_{n-1}))$$ (12')

where $s_1(r) = s(r) - r = s(w(r))$.

The above estimates will be called *aposteriori estimates* because the right hand side of (12') can be computed only *after* obtaining the points $x_1, \ldots, x_n$ via the iterative procedure. On the contrary the right hand side of (10) can be computed *before* obtaining these points. This is why the estimates (10) will be called *apriori* estimates. It is possible to prove that the aposteriori estimates are better than the apriori ones - at least in the case when w is nondecreasing, which is usually the case. More precisely we have:

1.10 <u>Proposition</u> *Suppose the hypotheses of Proposition 1.9 are satisfied and suppose that* $x_{n-1} \in Z(d(x_n, x_{n-1}))$ *for all* $n = 1,2,\ldots$. *If the rate of convergence* w *is nondecreasing then the following inequalities are satisfied:*

$$d(x_n, x^*) \leq s_1(d(x_n, x_{n-1})) \leq s(w^{(n)}(r_0)), \ n = 1,2,\ldots \tag{13}$$

where $s_1(r) = s(r) - r$.

<u>Proof</u> If the rate of convergence W is nondecreasing then the corresponding estimate function s will also be nondecreasing. From Proposition 1.9 we have $d(x_n, x_{n-1}) \leq w^{(n-1)}(r_0)$ so that

$$d(x_n, x^*) \leq s_1(d(x_n, x_{n-1})) = s(w(d(x_n, x_{n-1}))) \leq s(w(w^{(n-1)}(r_0)))$$

$$= s(w^{(n)}(r_0)). \qquad \square$$

In the following proposition we shall investigate the case when the above estimates are attained. This will turn out to be important later in the study of sharp estimates. A few words on the subject of sharpness of estimates are in order here. The word sharp is sometimes used in a rather loose way which might lead to misunderstandings; we shall give it a precise meaning later. For the moment, let us stress the fact that one of the advantages of the method of nondiscrete mathematical induction consists in the fact that it can be used to obtain sharp estimates for many important iterative procedures. Roughly speaking, a sharp estimate is one which cannot be improved. Here, however, we have to be careful what is meant by saying that an estimate cannot be improved.

If the estimate is given as a function of n characteristics, $a_1, \ldots, a_n$ say, of the object under consideration we shall say that it is sharp if there exists, within the class $C(a_1, \ldots, a_n)$ defined by the characteristics

9

$a_1...a_n$, at least one object for which the estimate is attained.

This does not exclude the possibility of obtaining a much better inequality if we have more information on the object to which we apply the iterative procedure.

Thus an estimate using the value of a function and of its derivative may be sharp even though better estimates may be obtained if we have information about higher derivatives as well.

1.11 <u>Proposition</u> *Suppose that – under the hypotheses of Proposition 1.9 equality is attained in the estimate (10) for a certain n_0. Then equality is attained in estimates (9) and (10) for all $n \geq n_0$.*

<u>Proof</u> From (9) of Proposition 1.9 we infer that $d(x_{k+1}, x_k) \leq w^{(k)}(r_0)$ for all $k \geq 0$. If (9) is verified with equality for $n = n_0$ then we may write

$$s(w^{(n_0)}(r_0)) = \sum_{k=n_0}^{\infty} w^{(k)}(r_0) = d(x_{n_0}, x^*)$$

$$\leq \sum_{n_0}^{n-1} d(x_k, x_{k+1}) + d(x_n, x^*) \leq \sum_{n_0}^{\infty} d(x_k, x_{k+1})$$

$$\leq \sum_{n_0}^{\infty} w^{(k)}(r_0) = s(w^{(n_0)}(r_0)).$$

It follows that $d(x_k, x_{k+1}) = w^{(k)}(r_0)$ for $k \geq n_0$ and that $d(x_n, x^*) = \sum_{n}^{\infty} w^{(k)}(r_0) = s(w^{(n)}(r_0))$ for each $n \geq n_0$. The proof is complete. □

Before we give significant applications, it will be interesting to clarify the relation of the induction theorem to two classical principles of functional analysis. We intend to show that it represents a generalization of the Banach fixed point theorem as well as of the closed graph theorem.

Relations to classical theorems

1.12 <u>The Banach Fixed Point Theorem</u> *Let (E,d) be a complete metric space and let f be a mapping of E into itself such that*

$$d(f(x), f(y)) \leq c\, d(x,y) \text{ for all } x, y \in E, \tag{14}$$

where c is a fixed number, $0 < c < 1$. Then:

(i)   f *has a unique fixed point* x^*;

(ii) *for any* $x_0 \in E$ *the sequence*

$$x_{n+1} = f(x_n), \quad n = 0,1,2,\ldots \tag{15}$$

converges to x^* *and the following estimates hold:*

$$d(x_n,x^*) \leq \frac{c^n}{1 - c}\, d(x_1,x_0), \quad n = 0,1,2,\ldots \tag{16}$$

$$d(x_n,x^*) \leq \frac{c}{1 - c}\, d(x_n,x_{n-1}), \quad n = 1,2,\ldots \,. \tag{17}$$

Proof   We shall apply Proposition 1.9 with the rate of convergence $w(t) = ct$ considered in Example 1.3 and the family of sets

$$Z(r) = \{x \in E;\ d(f(x),x) \leq r\}.$$

Let $x_0 \in E$ be given and set $r_0 = d(f(x_0),x_0)$. Condition (i) of Proposition 1.9 will be obviously satisfied. For any $x \in Z(r)$ set $x' = f(x)$. We have $d(x,x') \leq r$ and using (14) we may write

$$d(f(x'),x') = d(f(x'),\ f(x)) \leq c\, d(x',x) \leq c\, r = w(r).$$

The above inequality implies condition (ii) of Proposition 1.9. Consequently the sequence (x_n) will converge to a point $x^* \in Z(0)$ and the estimates (16) will be satisfied. The inclusion $x_{n-1} \in Z(d(x_n,x_{n-1}))$ being obvious the estimates (17) will also be satisfied. The fact that x^* is a fixed point of f follows immediately observing that (14) implies the continuity of f.

Let y^* be another point of E such that $y^* = f(y^*)$. According to (14) we have

$$d(x^*,y^*) = d(f^{(n)}(x^*),f^{(n)}(y^*)) \leq c\, d(f^{(n-1)}(x^*),f^{(n-1)}(y^*))$$

$$\leq \ldots \leq c^n\, d(x^*,y^*).$$

Since $0 < c < 1$ it follows that $x^* = y^*$. The proof is complete.   □

Examining the proof of the above theorem we observe that the concrete form of the rate of convergence w was not used in an essential manner. This remark allows us to state the following generalization of the Banach fixed point theorem:

1.13 Proposition *Let w be a nondecreasing rate of convergence on an interval T. Let E be a complete metric space and let f be a mapping of E into itself such that*

$$d(f(x), f(y)) \quad w(d(x,y)), \text{ for all } x,y \in E \tag{18}$$

Then:

(i) *for any $x_0 \in E$ the sequence $x_{n+1} = f(x_n)$ converges to a point $x^* \in E$ and the following estimates are satisfied*

$$d(x_n,x^*) \leq s(w^{(n)}(d(x_1,x_0))) \tag{19}$$

$$d(x_n,x^*) = s_1(d(x_n,x_{n-1})), \tag{20}$$

where s is the estimate function corresponding to w and $s_1(r) = s(r) - r$;

(ii) *if f is continuous then x^* is the unique fixed point of f.* □

1.14 Remark Point (i) of the above proposition holds under the weaker assumption

$$d(f^2(x), f(x)) \leq w(d(f(x),x)) \text{ for all } x \in E \tag{18'}$$

where $f^2(x) = f(f(x))$.

If $\lim_{r \downarrow 0} w(r) = 0$ then Condition (18) implies the continuity of f on E.

Now let us turn our attention to the closed graph theorem. We formulate it in its closed relation form [44], eliminating thereby all inessential assumptions which obscure its substance.

1.15 The Closed Graph Theorem *Let E be a complete metric space, and F a metric space. Let A be a closed subset of E × F. If the relation is uniformly almost open, then it is uniformly open.*

More precisely: suppose that, for each r > 0, there exists a positive number q(r) such that

$$(AU(x,r))^- \supset U(Ax,q(r))$$

for each $x \in D(A)$. *Then, for each* $r' > r$ *and each* $x \in D(A)$,

$$AU(x,r') \supset U(Ax,q(r)).$$

<u>Proof</u> Let $r > 0$, $r' > r$ and $x \in D(A)$ be fixed. Consider an arbitrary

$$y_0 \in U(Ax,q(r)).$$

We must show that $y_0 \in AU(x,r')$, or, in other words, that $A^{-1}y_0$ intersects $U(x,r')$. The proof is based on the following two observations.

It is not difficult to see that, if we replace the one point set y_0 by $U(y_0,t)$, it follows from the assumptions of the theorem that $A^{-1}U(y_0,t)$ does intersect $U(x,r)$ for arbitrarily small t. In fact, if we set

$$M(t) = A^{-1} U(y_0,t)$$

for $t > 0$, it is easy to infer from the assumption of the theorem that

$$M(q(t)) \subset U(M(s),t) \tag{21}$$

for arbitrarily small positive s. This means that, in a certain sense, the closed graph theorem is a limit case of the induction theorem; in fact, here we can take arbitrarily small functions $w(t)$.

The second observation is the following. It follows from the assumption that A is closed that $M(0) = A^{-1}y_0$. □   □

The above proof gives the closed graph theorem as a consequence of the induction theorem. In the following chapter we shall pursue this idea further and treat in greater detail a further generalization of the closed graph theorem in order to put into evidence the extremely simple general principle underlying all applications of the method.

Having cleared up how the induction theorem is related to classical results, let us now turn our attention to the principles of its application.

<u>Some principles of application of the induction theorem.</u>

Let us try to explain now why the method has been given the name of non-discrete mathematical induction.

Suppose we are given an approximation of order r, in other words a point $x \in M(r)$, and are allowed to move from x to a distance not greater than r.

Can we find, within $U(x,r)$, an approximation of a (much) better order r'? A suitable way of giving this a precise meaning is to impose the condition $r' = w(r)$, where w is a rate of convergence. The condition that for each $x \in M(r)$ there exists a point $x' \in U(x,r) \cap M(r')$ with $r' = w(r)$ may also be expressed in the form

$$M(r) \subset U(M(w(r)),r). \tag{22}$$

The possibility of passing from a given approximation $x \in M(r)$ to a better approximation $x' \in M(w(r))$ corresponds to the step from n to $n + 1$ in classical induction proofs.

If (22) is satisfied then the induction theorem asserts that $M(t) \subset U(M(0), s(t))$ for $t \in T$. Hence we shall be able to assert that $M(0)$ is nonvoid provided at least one $M(r)$ is nonvoid. Checking the fact that $M(r_0)$ is nonvoid for a certain $r_0 \in T$ corresponds to the first step in ordinary induction proofs; here, as in the discrete case, we have to make sure that the process begins somewhere.

There is another point which should be stressed: the heuristic value of the method.

The main advantage consists in the fact that the iterative construction is taken care of by the general theorem, so that the application consists in the verification of the hypotheses, the main question being: how much can a given approximation be improved within a given neighbourhood. By separating the hard analysis portion from the construction of the sequence, this method not only yields considerable simplifications of proofs but also brings out more clearly the substance of the problem. Instead of defining an approximation process first, and then investigating the degree of approximation at the n-th step, the method we propose could be described as exactly the opposite: we begin by looking at the sets $M(r)$ where the degree of approximation is at least r, then choose a suitable rate of convergence; the induction principle gives the construction of an iterative process corresponding to that rate of convergence automatically.

In this manner, we are using the relation between the improvement of the approximation and the distance we have to go in order to attain it in the most advantageous manner.

There are examples to show that a given system of functional inequalities may be consistent with different rates of convergence. The conclusion

obtained from the induction theorem may differ according to the choice of
these; however, there seems to be (at least in the concrete problems investigated thus far - in particular in the case of the Newton process, which we
shall discuss later) a natural rate of convergence which yields the best
possible result - in the sense that the estimates are sharp within the class
of problems under consideration.

Now let us try to formulate the above heuristic remarks somewhat more
precisely. Let (E,d) be a complete metric space and f a non-negative continuous function on E. We are looking for a point x for which $f(x) = 0$.

1st Observation Let us assume that, for each x taken from some set $D \subset E$ and
each positive $r < r_0$ we can prove an estimate of the form

$$\inf \{f(x'); x' \in U(x,r) \cap D\} < h(f(x),r)$$

where h is a suitable function of two variables. Suppose there exists a
positive function p tending to zero with r and a rate of convergence w such
that

$$h(p(r),r) \leq p(w(r)).$$

Set $M(r) = \{x \in D, f(x) \leq p(r)\}$; if h is increasing in the first argument
then $M(r) \subset U(M(w(r)),r)$.

Clearly the same conclusion remains true if h is only nondecreasing in
the first variable provided we replace the strict inequality in the above
estimate by $\leq$.

2nd Observation The functional equation connecting w and s may be used to
obtain information about the distance of the solution from any point u_0 given
in advance. Indeed, let u_0 be a fixed point in E. Given a point $x \in E$ and
two positive numbers d and r such that

$$d(x,u_0) \leq d - s(r),$$

then, for $x' \in U(x,r)$, we have

$$d(x',u_0) \leq d(x,u_0) + d(x',x) \leq d-s(r) + r = d-s(w(r)).$$

It follows that the family

$$Z(r) = \{x \in D; \ f(x) \le p(r), \ d(x,u_0) \le d-s(r)\}$$

satisfies $Z(r) \subset U(Z(w(r)),r)$. It follows that $Z(0)$ will be nonvoid if at least one $Z(r_0)$ is nonvoid since

$$Z(r_0) \subset U(Z(0), \ s(r_0)).$$

Summing up: if $h(p(r),r) \le p(w(r))$ and if there exists an $r_0 > 0$ and an $x_0 \in D$ such that

$$f(x_0) \le p(r_0) \ d(x_0,u_0) \le d-s(r_0)$$

then there exists an $x^* \in D^-$ with the following properties:

$$f(x^*) = 0,$$

$$d(x^*,x_0) \le s(r_0),$$

$$d(x^*,u_0) \le d.$$

In the above example we have looked for a function $h(m,r)$ with the following property: given x with $f(x) \le m$, there exists, within distance less than r, an x' with $f(x') \le h(m,r)$. In most cases the estimate for $f(x')$ will not depend on the value of $f(x)$ alone but will require some further characteristics as well; one might think of derivatives or some other additional information.

Suppose, for simplicity, that there is only one such additional characteristic, i.e. the estimate for $f(x')$ depends also on the value of another positive function f_1 at x so that $\inf \{f(x'), \ x' \in U(x,r) \cap D\} < h(f(x),f_1(x),r)$. Consider the case where the function h is increasing in the first argument and decreasing in the second argument. Since we shall need, in the following step of the induction, an estimate for $f_1(x')$, we shall need, in fact, a pair of positive functions h and h_1 such that, for each x and r, there exists an $x' \in U(x,r)$ for which

$$f(x') \le h(f(x), \ f_1(x),r),$$

$$f_1(x') \ge h_1(f(x), \ f_1(x),r).$$

In this case h_1 will have to be decreasing in the first argument and increasing in the second argument. It will then be desirable to find a pair of

16

functions p, p_1 and a rate of convergence w such that

$$h(p(r), p_1(r),r) \leq p(w(r)),$$

$$h_1(p(r),p_1(r),r) \geq p_1(w(r)).$$

Set $M(r) = \{x \in E;\ f(x) \leq p(r),\ f_1(x) \geq p_1(r)\}$; then $M(r) \subset U(M(w(r)),r)$.

   Let us pass now to examples which illustrate the general principles sketched above.

2 Three examples

In this section we intend to treat in detail three examples of the application of the method of nondiscrete mathematical induction. These examples happen also to be historically the first to be treated by this method. The third example is a fairly general nonlinear form of the closed graph theorem and is included in order to emphasize the extreme simplicity of the general principle underlying the method and to relate it to known theorems in functional analysis.

The first two examples are of a different character - they are intended to explain applications of the method to processes of numerical analysis and to iterative constructions in spectral theory. Two considerations were decisive in their choice: they should exhibit the essential features of the method without obscuring it by unnecessary technical details. We intend to explain carefully a way of finding an appropriate approximate set. There is, in general, more than one possibility of defining an approximate set corresponding in a natural manner to the problem under consideration and we shall see how their choice influences the result. Having then obtained the system of inequalities we shall turn our attention to its solution. We shall show that the most obvious choice of a solution does not necessarily lead to the best result for the original problem justifying thereby the introduction of more complicated rates of convergence. At the same time we shall provide the reader with two interesting examples of solutions of functional inequalities.

The reader who is only interested in the definitive results concerning the iterative procedures of numerical analysis may skip this section. He will find the complete statement of the theorems in Sections 4 and 5. But there less time will be devoted to explaining the motivation of the different steps of the proofs.

The main purpose of the present section is just to explain the method of the proofs.

The first example is the Newton process. There are two main reasons for starting with the Newton process besides the heuristic one mentioned above.

The first is the importance of the Newton process in the whole of Mathematics, the second is mainly historical, to give the reader the flavour of the early stages of the theory. We reproduce here essentially the first investigation of the Newton process based on the method of nmi (nondiscrete mathematical induction) as described in [50]. We do this in spite of the fact that the result obtained has been subsequently improved (as we shall see in Section 5) its heuristic value being decisive here.

The classical method of Newton for scalar valued functions is based on an inequality for the linear approximation of a function. Suppose f is a scalar valued function defined in a convex neighbourhood U of a point x_0 and suppose that it possesses in U a second derivative bounded by a constant k. For any, $x' \in U$ we have then

$$|f(x') - (f(x) + f'(x) (x' - x))| \leq \frac{1}{2} k |x' - x|^2. \tag{1}$$

It follows that, choosing x' so as to have

$$f(x) + f'(x) (x' - x) = 0,$$

we obtain

$$|f(x')| \leq \frac{1}{2} k |x' - x|^2. \tag{2}$$

This makes it possible to construct approximations for the solution of $f(x) = 0$. Indeed, if $\frac{f(x)}{f'(x)}$ is small, then, passing from the point x to the point $x' = x - \frac{f(x)}{f'(x)}$, we obtain a value f(x') which is small of the order $|\frac{f(x)}{f'(x)}|^2$.

The above inequality forms thus the essential basis of Newton's method. If properly interpreted, it extends to much more general situations, but nothing essentially different is needed to generalize the method to the case of Banach spaces.

We shall recall first some notions and simple facts to be used later.

If E is a Banach space we denote by $^d E$ its dual space and by B(E) the algebra of all bounded linear operators on E. The value of the functional $g \in {}^d E$ at the point x will be denoted either by g(x) or by $\langle x, g \rangle$. The latter notation which emphasizes the analogy to a scalar product is, sometimes, more convenient.

We shall also use a measure of invertibility for linear operators. If

$A \in B(E)$ we set

$$d(A) = \inf \{|Ax|; \; |x| \geq 1\};$$

clearly A is not invertible if $d(A) = 0$; also, even if $d(A)$ is positive, there will be instability of the inverse if $d(A)$ is small. Let us remind the reader of the fact that positivity of $d(A)$ does not imply invertibility of A in general (it does, though, in the finite dimensional case). However, if A is invertible then $d(A) > 0$ and $d(A) = |A^{-1}|^{-1}$.

The following perturbation properties of d are useful:

1. $d(A_2) \geq d(A_1) - |A_2 - A_1|$ *for all pairs* A_1, $A_2 \in B(E)$ $\hspace{2em}$ (3)

2. *if* A_1 *is invertible and* $|A_2 - A_1| < d(A_1)$ *then* A_2 *is invertible as well and*

$$|A_2^{-1}| \leq (d(A_1) - |A_2 - A_1|)^{-1}. \hspace{3em} (4)$$

If X and Y are Banach spaces, we denote by $B(X,Y)$ the linear space of all bounded linear operators from X into Y.

Let f be a (nonlinear in general) mapping defined on an open subset G of a Banach space X with values in a Banach space Y. Consider an $x_0 \in G$. We say that f is Fréchet differentiable at the point $x_0 \in G$ if there exists a linear operator $A \in B(X,Y)$ such that, for all sufficiently small $h \in X$

$$\lim \frac{1}{|h|} |f(x_0 + h) - f(x_0) - Ah| = 0 \quad \text{as} \quad |h| \to 0. \hspace{2em} (5)$$

It is easy to see that if such an operator A exists then it is unique. This operator A will be denoted by $f'(x_0)$ and will be called the Fréchet derivative of f at the point x_0. The reader will find more information about Fréchet derivatives in Appendix B.

At this moment it will be sufficient to see how to extend to the case of Banach spaces the fundamental inequality about quadratic approximation. It turns out that it is not necessary to assume the existence and boundedness of the second derivative: it suffices to have Lipschitz continuity of the first derivative. Indeed, the following lemma holds.

If the Fréchet derivative satisfies a Lipschitz condition

$$|f'(y) - f'(x)| \leq k|y - x| \hspace{4em} (6)$$

for all pairs y, x belonging to a segment $\{x_0 + t(x_1 - x_0), 0 \leq t \leq 1\}$ then

$$|f(x_1) - f(x_0) - f'(x_0) (x_1 - x_0)| \leq \frac{1}{2} k|x_1 - x_0|^2. \tag{7}$$

To prove this inequality, it suffices to write

$$f(x_1) - f(x_0) = \int_0^1 f'(x_0 + t(x_1 - x_0))dt \, (x_1 - x_0). \tag{8}$$

We have then

$$f(x_1) - f(x_0) - f'(x_0) (x_1 - x_0)$$

$$= \left(\int_0^1 f'(x_0 + t(x_1 - x_0))dt - f'(x_0) \right) (x_1 - x_0)$$

$$= \left(\int_0^1 (f'(x_0 + t(x_1 - x_0)) - f'(x_0))dt \right) (x_1 - x_0)$$

whence

$$|f(x_1) - f(x_0) - f'(x_0) (x_1 - x_0)|$$

$$\leq \int_0^1 kt|x_1 - x_0|dt \, |x_1 - x_0| = \frac{1}{2} k|x_1 - x_0|^2. \tag{9}$$

Let X and Y be two Banach spaces and let x_0 be a point of X. Let us denote by U the open sphere of X with centre x_0 and radius b and let us denote by $\bar{U}$ its closure. Let us consider a mapping $f:\bar{U} \to Y$ which is Fréchet different-iable on U and continuous on $\bar{U}$.

We are interested in giving sufficient conditions under which the follow-ing iterative procedure (Newton's method)

$$x_{n+1} = x_n - f'(x_n)^{-1} f(x_n) \tag{10}$$

is meaningful and produces a sequence (x_n) converging to a solution x^* of the equation $f(x) = 0$. We are also interested in estimating the errors $|x_n - x^*|$ at each step of the iterative procedure.

The iterative procedure is clearly of the form (1.7) with $Gx = x - f'(x)^{-1}f(x)$. Of course G will be defined on the set D of all points $x \in U$ for which $f'(x)$ is invertible.

Using the method of nondiscrete induction we shall prove that the follow-ing conditions are sufficient for the convergence of the Newton process

(G, x_0).

1. there exists a constant k such that

$$|f'(x) - f'(y)| \le k|x - y| \text{ for all } x, y \in U. \tag{11}$$

2. the linear operator $f'(x_0)$ is invertible and the following inequalities are satisfied

$$d(f'(x_0)) \ge d_0, \tag{12}$$

$$|f'(x_0)^{-1} f(x_0)| \le r_0, \tag{13}$$

$$2k \, r_0 \le d_0, \tag{14}$$

$$k^{-1}(d_0 - (d_0^2 - 2 k \, d_0 \, r_0)^{1/2}) < b. \tag{15}$$

This method will also yield error estimates of the form (1.10) and (1.12).

The convergence conditions as well as the error estimates obtained in this way will turn out to be optimal. For the precise meaning of optimality see Propositions 5.4 and 5.10.

Let us see now how the method of nondiscrete induction works:

We begin by defining an appropriate approximate set. Its definition must contain all the relevant information about an approximation and its limit must be the set of all solutions of $f(x) = 0$ (or contained therein). Given an approximation x, we expect

$$x' = x - f'(x)^{-1} f(x) \tag{16}$$

to be a better approximation; if r denotes the distance between x and x' we would like to have $x \in Z(r)$ and $x' \in Z(w(r))$ for some rate of convergence w. It will thus be useful to use the distance between x and x' as the parameter of the family and incorporate the condition $|f'(x)^{-1} f(x)| \le r$ in the definition of $Z(r)$. Another factor of decisive importance in passing from x to x' is the slope of the tangent (only figuratively speaking, as in the scalar case; in our case a tangent sufficiently steep is replaced by a Fréchet derivative whose invertibility $d(f'(x))$ is sufficiently high). We must take care that $d(f'(x))$ does not drop to zero and, indeed, that it does not get too small. Hence we impose a condition of the form $d(d'(x)) \ge h(r)$ for a positive function h to be determined later.

In the foregoing we have implicitly assumed that $Z(r)$ is contained in the domain of definition of f and f'. Nevertheless this is an important fact to take care of, so that in the definition of $Z(r)$ we should require the inequality

$$|x - x_0| \leq g(r),$$

where g is a function to be determined later, subject to the condition $g(r) < b$.

This leads to the choice

$$Z(r) = \{x \in X; \ |x - x_0| \leq g(r), \ f'(x) \text{ is invertible}, \tag{17}$$

$$|f'(x)^{-1} f(x)| \leq r, \ d(f'(x)) \geq h(r)\}, \ r > 0.$$

$Z(r)$ is obviously contained in the domain of definition of the mapping G, where $Gx = x - f'(x)^{-1} f(x)$.

Let us prove now that $x^* \in Z(0)$ implies $f(x^*) = 0$. From the definition of $Z(0)$ it follows that if $x^* \in Z(0)$ then there exist a sequence of positive numbers (r_n) with $\lim_{n \to \infty} r_n = 0$ and a sequence of points (x_n) such that $x_n \in Z(r_n)$ for all n, and $\lim_{n \to \infty} x_n = x^*$. Let us now consider a point $z \in Z(r)$ and write $x' = Gx$. Since $f(x') = f(x') - f(x) - f'(x) (x' - x)$ it follows from the Lipschitz continuity of f' that:

$$|f(x')| \leq \frac{1}{2} k |x' - x|^2 \leq \frac{1}{2} k \ r^2 \tag{18}$$

Taking in the above inequality $x = x_n$, $r = r_n$ and letting n tend to infinity we obtain that $\lim_{n \to \infty} f(x_n') = 0$. On the other hand we have

$$\lim_{n \to \infty} |x_n' - x^*| \leq \lim_{n \to \infty} r_n + \lim_{n \to \infty} |x_n - x^*| = 0 \tag{19}$$

and using the continuity of f on U we infer that $f(x^*) = 0$.

Let us show now that a triplet of functions g,h and w may be found such that the corresponding family $Z(\cdot)$ shall satisfy $Z(r) \subset U(Z(w(r)),r)$ for each $r > 0$.

Given $x \in Z(r)$, set $x' = Gx$ and let us try to find the smallest possible r' for which $x' \in Z(r')$. First we have to evaluate the distance $|x' - x_0|$.

Since $x \in Z(r)$ we have $|x' - x| \leq r$ and $|x - x_0| \leq g(r)$ so that we may write

$$|x' - x_0| \leq g(r) + r. \tag{20}$$

Next, we must take care of the invertibility of $f'(x')$. We have

$$d(f'(x')) = d(f'(x) + (f'(x') - f(x)))$$

$$\geq d(f'(x)) - |f'(x') - f'(x)| \geq d(f'(x)) - k|x' - x|,$$

and using the fact that $x \in Z(r)$ we obtain

$$d(f'(x')) \geq h(r) - kr. \tag{21}$$

As long as $h(r) - k r$ stays positive, we shall have invertibility of $f'(x')$ as well as the estimate

$$|f'(x')^{-1}| \leq (h(r) - k r)^{-1}.$$

This inequality together with (18) gives

$$|f'(x')^{-1} f(x')| \leq \frac{1}{2} k (h(r) - k r)^{-1} r^2. \tag{22}$$

Summing up, it follows that if x is an arbitrary point in $Z(r)$ and if $x' = Gx$ then the inequalities (20) - (22) are satisfied. We want to find a rate of convergence w such that $x' \in Z(w(r))$. In this case we shall have

$$x \in \bar{U} (Z(w(r)), r)$$

and, since x was arbitrary, the following inclusion will be satisfied

$$Z(r) \subset U(Z(w(r)), r).$$

This is actually the inclusion required in the induction theorem.

To have $x' \in Z(w(r))$ the following inequalities must hold:

$$|x' - x_0| \leq g(w(r)), \tag{23}$$

$$d(f'(x')) \geq h(w(r)), \tag{24}$$

$$|f'(x')^{-1} f(x')| \leq w(r). \tag{25}$$

We already have the estimates (20) - (22) for these quantities - using them we can obviously satisfy (23) - (25) by imposing the postulates:

$$g(r) + r \leq g(w(r)), \tag{26}$$

$$h(r) - k\,r \geq h(w(r)), \tag{27}$$

$$\frac{1}{2}\,k\,r^2\,(h(r) - k\,r)^{-1} \leq w(r). \tag{28}$$

Our task reduces thus to finding a rate of convergence w and two functions g and h which satisfy (26) - (28). The last two functions will be subject to the supplementary conditions $g(r) < b$ and $h(r) > 0$.

There are, in general, many solutions of the system of functional inequalities (26) - (28). Of course only those triplets w, g, h are of interest for which at least one corresponding $Z(r_0)$ is nonvoid. In this case, using the induction theorem, we infer the existence of an $x^* \in Z(0)$ for which

$$d(x^*, Z(r_0)) \leq s(r_0). \tag{29}$$

The inclusion $x^* \in Z(0)$ means that $f(x^*) = 0$, while relation (29) gives information as to the whereabouts of x^*. It is conceivable (and we shall see in the next example) that this information may be less precise if our choice of w, g, h, r_0 is not the most advantageous one. In the present case the most natural solution of the system of inequations happens to be also the optimal one.

Let us return to the system of inequations (26) - (28); writing m for the function defined by $h(t) = k\,m(t)$ our system simplifies to

$$g(r) + r \leq g(w\,(r)),$$

$$m(r) - r \geq m(w\,(r)),$$

$$\frac{1}{2}\,(m\,(r) - r)^{-1}\,r^2 \leq w(r).$$

Let us try to satisfy the first two inequations by imposing equality.

Recalling the fact that the estimate function s corresponding to a rate of convergence w satisfies the functional equation $s(r) = r + s(w(r))$, it follows that (26) and (27) will be satisfied with equality if we set $g(r) = c - s(r)$ and $m(r) = a + s(r)$ where c and a are two constants to be determined later.

The last inequation to be satisfied may then be rewritten in the form

$$\frac{1}{2} r^2 \leq w(r) \, m(w(r)).$$

Let us look for a solution of the corresponding equation. The following convention turns out to be useful. given a function v we shall denote by v' its composition with w, so that

$$v'(r) = v(w(r));$$

in particular, r' stands for w(r). Our system of functional equations assumes thus the following form:

$$g + r = g',$$

$$m - r = m',$$

$$\frac{1}{2} r^2 = r'm'.$$

A solution will satisfy

$$(m')^2 = (m - r)^2 = m^2 - 2mr + r^2 = m^2 - 2mr + 2m'r',$$

whence

$$m'^2 - 2m'r' = m^2 - 2mr.$$

The function $m^2 - 2mr$ assumes thus the same value at the points r and w(r). It is thus natural to look for solutions for which $m^2 - 2mr$ is a constant. Letting r tend to zero we shall see that this constant must be a^2, so that

$$m^2 - 2mr - a^2 = 0.$$

It will follow that $m(r) = r + (r^2 + a^2)^{1/2}$, whence

$$s(r) = r + (r^2 + s^2)^{1/2} - a, \tag{30}$$

$$w(r) = \frac{1}{2} r^2 (r^2 + a^2)^{-1/2} \tag{31}$$

The following reasoning justifies this conclusion. Suppose $m^2 - 2mr$ is a constant, q say. Since $m(r) \geq r$ it follows that $m(r) = r + (r^2 + q)^{\frac{1}{2}}$ and thus q must be non-negative. Now $s(r) = m(r) - a = r + (r^2 + q)^{\frac{1}{2}} - a$,

$w(r) = \frac{1}{2} r^2(r^2 + q)^{-\frac{1}{2}}$. Since $s(r) - r = m(r)-r-a = m(w(r)) - a = s(w(r))$ we wee that w is indeed a rate of convergence by Proposition (1.2). Being its estimate function, s tends to zero as $r \to 0$ and it follows that $q = a^2$.

We have thus obtained a whole family of solutions w, g, h of the system (26) - (28), depending on the parameters a and c. Let us choose these parameters to our best advantage.

The natural requirement which still has to be satisfied is that $x_0 \in Z(r_0)$. For this it is necessary that $d(f'(x_0)) \geq h(r_0)$. According to (12) the last inequality will be satisfied if

$$d_0 \geq h(r_0) = k(r_0 + (r_0^2 + a^2)^{1/2}).$$

(Observe that $h(r_0) \geq 2kr_0$). It is obvious that the largest possible a will give the best result. This leads to the choice

$$\left(\frac{d_0}{k} - r_0\right)^2 = r_0^2 + a^2$$

$$a^2 = \left(\frac{d_0}{k}\right) - 2 \frac{d_0}{k} r_0 = \frac{d_0}{k^2} (d_0 - 2 k r_0).$$

The application of our method required in a natural manner the condition $d_0 \geq 2 k r_0$. At the moment this appears as a requirement imposed by the method of the proof - we shall see later that it is also a natural condition for the convergence of the process itself.

Finally we have to determine the parameter c in such a way that $x_0 \in Z(r_0)$ and $g(r) < b$ for all $r > 0$. These requirements will be satisfied if we have $b > c \geq s(r_0)$. The most advantageous way of choosing c is, of course, $c = s(r_0)$. Since $s(r_0) = k^{-1}d_0 - a$ the inequality $b > c$ reduces to condition (15).

With the above definitions of $Z(\cdot)$ and w, the hypotheses (i) and (ii) of Proposition 1.9 are satisfied. Consequently it follows that the iterative procedure (10) is meaningful and that the sequence (x_n) produced by it converges to a root x^* of the equation $f(x) = 0$. It also follows that for each $n = 0,1,...$ estimates of the form

$$|x_n - x^*| \leq s(w^{(n)}(r_0)) \tag{32}$$

are satisfied.

We can obtain better estimates using point 3 of Proposition 1.9. For this let us first prove that

$$x_{n-1} \in Z(|x_n - x_{n-1}|) \tag{33}$$

for $n = 1,2,\ldots$. From point 2 of Proposition 1.9 we already know that $x_{n-1} \in Z(w^{(n-1)}(r_0))$.

In particular we have $|x_n - x_{n-1}| \leq w^{(n-1)}(r_0)$; s being an increasing function it follows that $s(|x_n - x_{n-1}|) \leq s(w^{(n-1)}(r_0))$. Consequently, if $x_{n-1} \in Z(w^{(n-1)}(r_0))$ we may write

$$|x_{n-1} - x_0| \leq g(w^{(n-1)}(r_0)) = c - s(w^{(n-1)}(r_0))$$

$$\leq c - s(|x_n - x_{n-1}|) = g(|x_n - x_{n-1}|)$$

$$d(f'(x_{n-1})) \geq h(w^{(n-1)}(r_0)) = k(a+s(w^{(n-1)}(r_0))$$

$$\geq k(a + s(|x_n - x_{n-1}|)) = h(|x_n - x_{n-1}|).$$

This shows that (33) is satisfied and according to point 3 of Proposition 1.9 we obtain the estimate

$$|x_n - x^*| \leq (a^2 + |x_n - x_{n-1}|^2)^{1/2} - a. \tag{34}$$

In Section 5 we shall reconsider the study of Newton's process, improving the results obtained above. Namely, instead of conditions (12) - (14) we shall require the weaker conditions

$$|f'(x_0)^{-1} (f'(x) - f'(y))| \leq k_0 |x - y| \tag{35}$$

$$2 k_0 r_0 \leq 1. \tag{36}$$

It turns out that these conditions are "affine invariant". The importance of having such conditions is stressed in [11]. We shall also give an explicit form to the quantities $s(w^{(n)}(r_0))$ appearing in (32).

Our second example is the iterative procedure treated first in the Gatlinburg lecture [49]. It is, historically, the first example of an application of the method of nondiscrete mathematical induction where the necessity of using nonlinear rates of convergence appears in a natural manner. We shall attempt to reproduce here all the stages of the process through which, by trial and error, we arrive at the correct solution of the given problem.

We shall first set up a system of inequalities corresponding to what seems to be the natural approximate set for the given problem; as a first try, we look for linear solutions of the inequalities - the induction theorem yields then a result for our problem, an estimate for the spectrum of an operator. This estimate, however, turns out to be sharp only asymptotically. Replacing the linear rate of convergence by the "correct" one, we obtain an estimate which is sharp. The correct rate of convergence is asymptotically linear so that it might seem, at first glance, a small gain at the expense of a more complicated formula for the estimates of the consecutive steps of the iteration. Using the linear rate of convergence means, however, a loss in the estimates at each step. These differences, minute as they might be numerically, nevertheless add up to a non-negligible loss of information in the final estimate.

The problem to be considered is as follows. Suppose we have a matrix

$$A = \begin{pmatrix} a_{11}, & a_{12}, \ldots, & a_{1n} \\ a_{21}, & a_{22}, \ldots, & a_{2n} \\ a_{n1}, & a_{n2}, \ldots & a_{nn} \end{pmatrix} .$$

Denote by u and v the vectors of length n-1 defined by

$$u^T = (a_{21}, \quad a_{31}, \ldots, \quad a_{n1})$$

$$v = (a_{12}, \quad a_{13}, \ldots, \quad a_{1n}) .$$

If one of the vectors u or v is very small, it might be expected that A will have an eigenvalue close to a_{11}. Our task will be to give this statement a precise formulation and give the best possible estimate for the distance of that eigenvalue from a_{11}. We shall treat, again, the case of an operator in a Banach space.

Before starting let us state a simple formula which will be used frequently in the sequel. Consider two operators $A_1, A_2 \in B(E)$, where E is a Banach space.

If A_1 and A_2 are both invertible then

$$A_2^{-1} - A_1^{-1} = A_2^{-1}(A_1 - A_2)A_1^{-1}.$$

Suppose the space E is written as the direct sum $E = E_1 \oplus E_2$ where E_1 is one-dimensional. Given $u \in E_2$, $v \in {}^dE_2$, $C \in B(E_2)$ and a_{11} a complex number we write

$$\begin{pmatrix} a_{11} & v \\ u & C \end{pmatrix}$$

for the operator A defined on E as follows: the equation $y = Ax$ is equivalent to

$$y_1 = x_1 a_{11} + \langle x_2, v\rangle$$

$$y_2 = x_1 u + Cx_2$$

Now suppose that C^{-1} exists. Upon postmultiplying A by the (clearly non-singular) operator

$$\begin{pmatrix} 1 & 0 \\ -C^{-1}u & I \end{pmatrix}$$

we have

$$\begin{pmatrix} a_{11} & v \\ u & C \end{pmatrix}\begin{pmatrix} 1 & 0 \\ -C^{-1}u, & I \end{pmatrix} = \begin{pmatrix} a_{11} - \langle C^{-1}u,v\rangle & v \\ 0 & C \end{pmatrix}.$$

This immediately gives the following proposition which will be used later.

If C^{-1} exists the operator A is nonsingular if and only if $a_{11} - \langle C^{-1}u,v\rangle \neq 0$.

Let us recapitulate briefly the statement of the problem. Let A be an operator of the form

$$A = \begin{pmatrix} a_{11} & v \\ u & C \end{pmatrix}$$

with $C - a_{11}$ invertible; if u and/or v is small it is to be expected that, close to a_{11}, there will be a point of the spectrum of A. Set $a = |u| \; |v|$ and $b = d(C - a_{11})$. Our task consists in *finding the smallest possible number r (in terms of a and b) such that the disc $\{z; |z - a_{11}| \leq r\}$ will intersect the spectrum of* A.

The condition for A - λ to be singular reads $a_{11} - \lambda - \langle(C-\lambda)^{-1}u,v\rangle = 0$. Since $C - a_{11}$ is invertible $(C - \lambda)^{-1}$ will exist if λ is close to a_{11}. If we set $z = \lambda - a_{11}$ and $B = C - a_{11}$ the problem reduces to finding a small z with $-z - \langle(B - z)^{-1}u,v\rangle = 0$.

Let G be the set of all complex z for which $(B - z)^{-1}$ exists so that G contains all z less than $b = d(B)$ in modulus. For $z \in G$ set

$$g(z) = \langle(B - z)^{-1}u,v\rangle. \tag{37}$$

Conditions are to be found which will ensure the existence of a z for which $z + g(z) = 0$. For $r > 0$ set

$$M(r) = \{z; |z + g(z)| \leq r, d(B - z) \geq h(r)\} \tag{38}$$

where h is a *positive* function to be chosen later. Now suppose that $z \in M(r)$ and set $z' = -g(z)$. It follows that $|z' - z| = |g(z) + z| \leq r$ whence $d(B - z') \geq d(B - z) - |z' - z| \geq h(r) - r$. If $h(r) - r$ is positive, $z' \in G$ and $g(z')$ is defined. We then have

$$z' - z = -(z + g(z)),$$

$$z' + g(z') = g(z') - g(z) = (z'-z)\langle(B-z)^{-1}(B - z')^{-1}u,v\rangle,$$

so that $|z' - z| \leq r$ and $|z' + g(z')| \leq r\,h(r)^{-1}(h(r) - r)^{-1}a$.

If we can find a rate of convergence w such that these inequalities imply $z' \in M(w(r))$ we shall have $z' \in M(w(r)) \cap \bar{U}(z,r)$, in other words, the inclusion

$$M(r) \subset \bar{U}(M(w(r)),r).$$

For the inclusion $z' \in M(w(r))$ the following two inequalities will clearly be sufficient

$$ar(h(r)\cdot(h(r) - r))^{-1} \leq w(r) \tag{39}$$

$$h(r) - r \geq h(w(r)). \tag{40}$$

Notice that the second inequality ensures, in particular, that $g(z')$ is meaningful. In order to apply the induction theorem, we need an $r_0 > 0$ such that $M(r_0)$ is nonvoid; this requirement will be realized by the inclusion

$0 \in M(r_0)$. The induction theorem says that, for such r_0,

$$M(r_0) \subset \bar{U}(M(0), s(r_0))$$

so that there exists a $z \in M(0)$ for which $|z| \leq s(r_0)$; clearly $z \in M(0)$ implies $z + g(z) = 0$ so that $\lambda = z + a_{11}$ belongs to the spectrum of A. (Of course, to justify this conclusion we must know that $z \in G$; this will be guaranteed by the inequality $s(r_0) < b$).

In other words, if we find a rate of convergence w and a positive function h which satisfy the above functional inequalities and if $0 \in M(r_0)$ then the disc $\{\lambda; |\lambda - a_{11}| \leq s(r_0)\}$ will intersect the spectrum of A.

In our particular case the inclusion $0 \in M(r_0)$ is equivalent to $|g(0)| \leq r_0$ and $b = d(B) \geq h(r_0)$.

The problem reduces thus to the following: *to find two positive functions* w *and* h *defined on an interval* T *such that*

> w *is a rate of convergence on* T $\qquad\qquad\qquad$ (41)

> $ar \leq w(r) \cdot h(r) \cdot (h(r) - r) \qquad \forall r \in T$ $\qquad\qquad$ (42)

> $h(r) - r \geq h(w(r)) \qquad\qquad \forall r \in T$ $\qquad\qquad$ (43)

> *there exists an* $r_0 \in T$ *such that* $|g(0)| \leq r_0$
> *and* $d(B) \geq h(r_0)$. $\qquad\qquad\qquad\qquad\qquad\qquad\qquad$ (44)

As in Definition 1.1, T has to be either an interval of the form $\{t; 0 < t \leq t_0\}$ or the set of all positive numbers. In our case we shall have $T = \{t; t > 0\}$.

If w is a rate of convergence and s the corresponding estimate function, then, for any $p \in \mathbb{R}$, the function $h(r) = p + s(r)$ satisfies the functional equation $h(r) - r = h(w(r))$. In particular h will satisfy (43). Since h is to be positive, p must be positive.

It is conceivable that, for a suitable choice of p and w, the function $h(r) = p + s(r)$ will satisfy also the requirement (42). Let us examine the meaning of condition (44) for this choice of h. This condition may be written in the form $s(r_0) \leq b - p$ where $b = d(B)$ and $r_0 \geq |g(0)|$.

At this stage the first possibility which offers itself is to try to satisfy our system of functional inequalities by a linear rate of convergence. Take w in the form $w(r) = cr$ with c to be chosen later; it follows that (see

Example 1.3.)

$$h(r) = p + \frac{1}{1 - c} \, r,$$

so that, in this particular case, the conditions to be satisfied are as follows:

$$0 < c < 1, \tag{41'}$$

$$a \leq c(p + \frac{1}{1 - c} \, r) \, (p + \frac{c}{1 - c} \, r), \tag{42'}$$

$$\frac{1}{1 - c} \, r_0 \leq b - p, \quad r_0 = |g(0)| \tag{44'}$$

To satisfy (42'), it suffices to have $a \leq cp^2$; an obvious choice - which turns out to be feasible - is to postulate $a = cp^2$. Hence we set $p = (a/c)^{1/2}$, reducing thereby the number of parameters to be chosen to one.

Now condition (44') cannot be satisfied unless $b - (a/c)^{1/2} = b - p \geq 0$; it follows that it will be necessary to assume $b^2 > a$. Since $|g(0)| \leq a/b$ condition (44') will be satisfied if

$$a/b \leq (1 - c) \, (b - (a/c)^{1/2})$$

$$1 \leq (1 - c) \, (b^2/a - (b^2/a \, c)^{1/2}).$$

Now introduce a new parameter k by the equation $k = (b^2c/a)^{1/2}$ so that $(b^2/ac)^{1/2} = b^2/ak$; to satisfy the above inequality, we must require k to be greater than one. In terms of k, the inequality to be satisfied becomes

$$b^2/a \geq k^2 + k/(k - 1). \tag{45}$$

Let us denote by k_0 the (unique) point $k_0 > 1$ where the function $k^2 + k/(k-1)$, $k > 1$ assumes its minimum m_0. Let us show now that it suffices to postulate $b^2/a \geq m_0$ and set $c = (a/b^2)k_0^2$. Indeed, if $b^2/a \geq m_0$, we have

$$1 \geq \frac{a}{b^2} k_0^2 + \frac{a}{b^2} \frac{k_0}{k_0 - 1} , \tag{46}$$

so that $c = (a/b^2)k_0^2 < 1$. For this c, the rate of convergence $w(r) = cr$ satisfies all our requirements, so that the induction theorem applies. We may thus formulate the following result:

If $b^2 \geq m_0 a$, *then there exists a point* λ *in the spectrum of* A *with*

$$|\lambda - a_{11}| \leq \frac{1}{1-(a/b^2)k_0^2}| \quad |<B^{-1}u,v>| \leq \frac{a}{b-(a/b)k_0^2} .$$

For m_0, k_0 *and* k_0^2 *the following approximate values may be obtained*

$$m_0 = 5.22, \quad k_0 = 1.56, \quad k_0^2 = 2.44.$$

This seems to be as much as may be obtained using a linear rate of convergence.

By using a slightly more refined rate of convergence we will be able, in what follows, to obtain the exact result.

To this end let us return to the system of functional inequalities (42) - (43). We have seen in the foregoing discussion that each solution of this system defines a family

$$M(r) = M(w,h;r) = \{z; |z + g(z)| \leq r, d(B-z) \geq h(r)\} \tag{47}$$

and yields an existence theorem provided at least one M(r) is nonvoid. In general, there is no need to find all solutions of the system (41) - (44); it is enough to find one for which the corresponding existence theorem is sharp.

In our case it turns out that it suffices to look for w and h for which equality is attained. We start with a given a > 0 and look for pairs w, h - (w a rate of convergence, h positive) - which satisfy

$$h(r) - r = h(w(r)) \tag{48}$$

$$ar = w(r) h(r) (h(r) - r) \tag{49}$$

for all $r \in T$. Write, for shortness, $r' = w(r)$ and $h' = h(r')$. Any solution of the above system satisfies $r = h - h'$, whence

$$r' = ar/(h\ h') = a(h - h')/(h\ h') = a(1/h' - 1/h)$$

so that

$$a/h' - a/h + h' - h = r' - r.$$

It follows that

$$a/h' + h' - r' = a/h + h - r.$$

Thus we have to look for solutions of our system among the functions h for which $a/h(r) + h(r) - r$ is a constant, say β. Such functions satisfy the equation $h^2 - (\beta + r)h + a = 0$.

Since h is to be positive the discriminant must be non-negative for all $r > 0$; the condition $\beta^2 \geq 4a$ must be imposed. For such pairs a,β

$$h(r) = \frac{1}{2}(\beta + r + (\beta + r)^2 - 4a)^{1/2}) \tag{50}$$

turns out to be a solution of our problem together with

$$w(r) = \frac{ar}{h(r)\ (h(r)-r)} = r\ \frac{\beta + r - u}{\beta - r + u} \tag{51}$$

where $u(r) = ((\beta + r)^2 - 4a)^{1/2}$. According to Proposition 1.2 it follows from (48) that w is a rate of convergence on T and that the corresponding estimate function is given by

$$s(r) = h(r) - h(0) = h(r) - \frac{1}{2}(\beta + u_0) = \frac{1}{2}(r + u(r)) - \frac{1}{2}u_0. \tag{52}$$

Here we write u_0 for $u(0) = (\beta^2 - 4a)^{1/2}$.

It is interesting to observe that the function w satisfies a quadratic equation. Writing w under the form

$$w = r\ \frac{\beta - (u - r)}{\beta + (u - r)}$$

and solving for $u - r$ we have

$$u - r = \beta\ \frac{r - w}{r + w}.$$

The above equation is, in fact, a quadratic equation for w, since it may be rewritten as

$$(r + \beta\ \frac{r - w}{r + w})^2 = (\beta + r)^2 - 4a. \tag{53}$$

This fact may be used for the direct verification of the relation

$$s(w(r)) + r = s(r) \text{ for } s(r) = \frac{1}{2}(r + u(r)) - \frac{1}{2}u_0,$$

thereby giving an alternative derivation of the solution.

We shall adopt the convention that has already proved itself useful above: if f is a function, we write f' for the composition of f and w. Thus f'(r) = f(w(r)). In particular r' = w(r). The relation to be verified is s' + r = s, in other words $\frac{1}{2}$ (r + u)' + r = $\frac{1}{2}$ (r + u). This can be written as

$$u' = u - r - w. \tag{54}$$

Using relation (53) it is not difficult to verify (54) as follows

$$(u - r - w)^2 = u^2 - 2u(r + w) + (r + w)^2$$

$$= u^2 - 2(u - r) \cdot (r + w) - 2r (r + w) + (r + w)^2$$

$$= u^2 - 2\beta (r - w) - (r + w) (r - w)$$

$$= r^2 + 2r\beta + \beta^2 - 4a - 2\beta (r - w) - r^2 + w^2$$

$$= \beta^2 + 2\beta w + w^2 - 4a = (u')^2.$$

Up to this point the functions w and h have been investigated as solutions of a system of functional inequalities, without regard to their interpretation as estimates for certain quantities occurring in an existence problem. Although no attempt was made to find all solutions of (42) - (43) we still have, for each $\beta^2 \geq 4a$, a pair w and h satisfying the inequalities (42) - (43). To each such a pair there corresponds a family of sets M(r) which satisfies the induction relation

$$M(r) \subset \bar{U}(M(w(r)),r).$$

To obtain an existence theorem we need an r_0 for which the corresponding $M(r_0)$ is nonvoid. As in the preceding case we shall realize this by proving that there exists an r_0 for which $0 \in M(r_0)$. This is equivalent to (44). We have first $|g(0)| = |\langle B^{-1}u,v\rangle| \leq a/d(B)$ where a = $|u|$ $|v|$. The constant b which defines h is still to be chosen. It is easy to check that h(a/β) =β. Hence if we set β = d(B), and assume d(B)$^2 \geq 4|u|$ $|v|$, we shall have $|g(0)| \leq a/\beta$ and h($|g(0)|$) $\leq$ h(a/β) = d(B), so that $|g(0)|$ can be taken for r_0. Then s(r_0) may be estimated as follows (writing b for β)

36

$$s(r_0) \leq s(a/b) = h(a/b) - h(0) = \frac{1}{2}(b - (b^2 - 4a)^{1/2}).$$

This completes all the information necessary for the application of the induction theorem. Consequently we obtain the following result.

If $b^2 \geq 4a$, then there exists a point λ in the spectrum of A such that

$$|\lambda - a_{11}| \leq s(|<B^{-1}u,v>|) \leq \frac{1}{2}(b - (b^2 - 4a)^{1/2}). \tag{55}$$

This result turns out to be *sharp*; for the matrix

$$\begin{pmatrix} 0 & a^{1/2} \\ -a^{1/2} & b \end{pmatrix}$$

one of the eigenvalues lies exactly on the circumference of the disk $\{t; |z| \leq \frac{1}{2}(b - u_0)\}$. The result established above improves the one obtained with the linear rate of convergence in two directions. First of all the condition $b^2 \geq m_0 a$ imposed there is stronger than the condition $b^2 \geq 4a$, because $m_0 \approx 5.22$. Second, the estimates (55) are better than the estimates (47). However, they are not much more so. Indeed, if we write

$$s_1(a,b) = \frac{a}{b - (a/b)k_0^2} \quad \text{and} \quad s_2(a,b) = \frac{1}{2}(b - (b^2 - 4a)^{1/2})$$

it may be proved that

$$s_2(a,b) < s_1(a,b) \leq \frac{3-k_0}{2k_0(2-k_0)} s_2(a,s) < 1.06 s_2(a,b)$$

for all $a > 0$, $b > 0$ with $b^2 \geq 4a$.

To illustrate the possibilities of the method of nondiscrete mathematical induction let us conclude this section by remarking that, in the original paper [49], the functions w and s from (50) and (51) were obtained in a different manner, based on yet another use of the functional equation relating w and s.

It is a well known fact that, in proofs using the classical induction method, it is sometimes more advantageous to prove a statement stronger than the required one, because - at the induction step - the induction hypothesis contains more information which can be used with advantage to prove the next step of the induction. This rather trivial observation turns out to be true in the nondiscrete case as well. Let us illustrate this by reproducing here the original proof from [49] where a more restrictive approximate set was used.

By imposing further restrictions on the family $M(\cdot)$ we may, sometimes, obtain a more precise result or a simpler proof.

Let us begin by observing that we have the following estimates

$$d(B - z') \geq d(B) - |z'| = d(B) - |g(z)|$$

$$|g(z)| \leq \frac{a}{h(r)}$$

whence

$$d(B - z') \geq b - \frac{a}{h(r)} \quad .$$

It is thus natural to impose the following additional requirement

$$b - \frac{a}{h(r)} \geq h(w(r)) \tag{56}$$

and define an approximate set as follows

$$Z(r) = \{z; |z + g(z)| \leq r, \ d(B - z) \geq h(r), \ b - |z| \geq h(r)\}.$$

If conditions (41), (42), (43) and (56) are satisfied, the inclusion $Z(r) \subset U(Z(w(r)),r)$ will hold. As in the preceding example we shall satisfy (43) by postulating equality. To satisfy (56), it will be sufficient to have

$$h - r - b + \frac{a}{h} = 0$$

or

$$h(h - r - b) + a = 0. \tag{57}$$

In order to ensure the existence of a solution of (57) for small positive r it will be necessary to impose the condition $b^2 \geq 4a$. Having done this it is easy to see that, in order to satisfy (57), it will be sufficient to set

$$h(r) = \frac{b + r}{2} + \left((\frac{b + r}{2})^2 - a \right)^{1/2}$$

(since $h(r) = m + s(r)$ for some rate of convergence w, h has to satisfy the inequality $h(r) \geq m + r$, hence the plus sign in the solution of the quadratic equation). In order to obtain the rate of convergence w corresponding to the function $s = h - m$ it is necessary to compute the inverse function s^{-1}. If

38

$y > 0$ is given, the value $s^{-1}(y)$ is the solution of the equation $s(r) = y$. Now $y = s(r)$ implies

$$(y + m)^2 - (r + b)(y + m) = -a$$

whence

$$r + b = \frac{(y + m)^2 + a}{y + m} \ .$$

It follows that

$$w(r) + b = \frac{(y' + m)^2 + a}{y' + m}$$

for $y' = s(r) - r$. Since $y' + m = m + s(r) - r = h(r) - r$, we obtain, using (57)

$$w(r) = r \, \frac{b + r - h}{h-r} \ .$$

Using the abbreviation $u = ((b + r)^2 - 4a)^{1/2}$ we have $h = \frac{1}{2}(b + r + u)$ whence

$$w(r) = r \, \frac{b + r - u}{b - r + u} \ .$$

It is easy to check the inequality

$$ar \leqq w(r) \, h(h - r) \tag{42}$$

since

$$a \leqq \frac{b + r - h}{h-r} \, h(h - r)$$

by (57). Indeed, it follows from (57) that

$$ar = r(b + r - h)h = r \, \frac{b + r - h}{h-r} \, h(h-r) = w(r) \, h(h-r).$$

Also, it is not difficult to prove that $0 \in Z(a/b)$.

Our third example is a nonlinear generalization of the subtraction theorem and its purpose is twofold: first it shows the origin of the method, second it serves also to illustrate the general idea linking up results belonging to functional analysis to the study of iterative processes of numerical mathematics.

The reader is strongly advised to acquaint himself with this third example as well, but the knowledge of it is not indispensable for the understanding of the material treated in Sections 7-11. These sections are self-contained and may be read independently but most likely the reading will be less easy for those who did not work their way through the present section.

In 1965 the second author observed [43] that the classical closed graph theorem may be extended to a more general result which has a quantitative character.

The juxtaposition of the qualitative and quantitative forms of the classical closed graph theorem which follows might help to clarify the distinction between the two. If E and F are two normed linear spaces with closed unit cells U and V, respectively, u a bounded linear mapping of E into F, the classical open mapping theorem may be formulated as follows:

2.1 <u>Theorem</u> *Suppose* $u(U)^- \supset \beta V$ *for some* $\beta > 0$. *If* E *is complete then*

$$u(U) \supset (\beta - \epsilon)V$$

for each $0 < \epsilon < \beta$.

In particular, the mapping u is open.

The assumption $u(U)^- \supset \beta V$ means that every element of βV may be arbitrarily well approximated by elements of $u(U)$. This is, indeed, much more than is necessary: the openness of the mapping may be salvaged even if the approximation is less good. The distance of each element of βV to $u(U)$ does not have to be zero, it suffices if it is smaller than β. This observation was the starting point of the paper [43] where the following subtraction theorem was proved.

2.2 <u>Theorem</u> *Suppose* $u(U) + \alpha V \supset \beta V$ *for some* $0 < \alpha < \beta$. *If* E *is complete then*

$$u(U) \supset (\beta - \alpha)V.$$

We observe that we still have openness of the mapping u except that now we can only assert that $u(U)$ contains a cell of a smaller radius $\beta - \alpha$ as opposed to the preceding case where the radius could be taken arbitrarily close to β.

40

It is not difficult to see that linearity is not essential in this result. Indeed, it was just this observation which set off the development of the method of nondiscrete mathematical induction. In this section we prove what seems to be the natural nonlinear generalization of the subtraction Theorem 2.2; not only does it contain Theorem 2.2 as a very particular case but also the proof represents a considerable simplification of the original one.

In the assumptions of the linear subtraction theorem the condition $\alpha < \beta$ is of decisive importance: the degree of approximation must be sufficiently good as compared with the measure of openness β. The condition $\alpha < \beta$ may be restated in the following equivalent form: if we consider the functions

$$m(t) = \alpha t$$

$$q(t) = \beta t$$

then the condition $\alpha < \beta$ is equivalent to the requirement that $q^{(-1)} \circ m$ be a rate of convergence. This seems to be the right interpretation of the relation between α and β: this conjecture is borne out by the fact that the theorem - when restated in terms of m and q - remains true in the nonlinear case as well: see Theorems 2.3 and 2.4 of this section.

Given two sets E and F, a relation in $E \times F$ is defined as a subset of $E \times F$. The points of the Cartesian product $E \times F$ will be written as $[x,y]$ where $x \in E$ and $y \in F$. If S is a relation in $E \times F$ then the inclusion $[x,y] \in S$ will, sometimes, be written in the form xSy. Given $x \in E$ we define

$$Sx = \{v \in F; \ xSv\}.$$

Given $y \in F$ we set

$$S^{-1}y = \{u \in E; \ uSy\}.$$

Hence the inclusion $[x,y] \in S$ may also be written - whenever it is convenient - in each of the following equivalent forms

$$xSy, \quad x \in S^{-1}y, \quad y \in Sx.$$

If $A \subset E$ we set

$$SA = \{v \in F; \ S^{-1}v \cap A \neq \emptyset\}$$

so that

$$SA = \bigcup_{a \in A} Sa.$$

In a similar manner, for $B \subset F$ we define

$$S^{-1}B = \{u \in E, \; Su \cap B \neq \emptyset\} = \bigcup_{b \in B} S^{-1}b.$$

Since the generalization of the subtraction theorem we intend to prove is not immediately recognizable as such, we shall state first a weaker version of which Theorem 2.2 is an immediate corollary.

2.3 __Theorem__ *Let E and F be two metric spaces and let a closed relation $S \subset E \times F$ be given. Suppose m and q are two positive functions defined on T such that*

$$U(SU(x,r), \; m(r)) \supset U(Sx, q(r))$$

for all $x \in E$ and all $r \in T$.

   Assume further that:

1. *m is small with respect to q,   i.e. there exists a rate of convergence w on T such that $m(t) \leq q(w(t))$ for all $t \in T$;*

2. *$m(t) \to 0$ as $t \to 0$;*

3. *the space E is complete.*

   Then

$$SU(x, \; s(r)) \supset U(Sx, q(r))$$

for all $x \in E$ and all $r \in T$.

   Let us show now that Theorem 2.2 is an immediate consequence of Theorem 2.3. Indeed, suppose the assumptions of Theorem 2.2 satisfied. It is not difficult to verify that the condition of Theorem 2.3 are fulfilled for the following choice of the functions m, q, w

$$m(t) = \alpha t, \; q(t) = \beta t, \; w(t) = \frac{\alpha}{\beta} t;$$

the assumption $\alpha < \beta$ guarantees that w is a rate of convergence. Note that

$$\sigma(t) = \frac{1}{1 - \frac{\alpha}{\beta}} \, t .$$

Now it suffices to write down the assertion of Theorem 2.3 for $x = 0$ and $r = 1$; this gives

$$S(\frac{1}{1 - \frac{\alpha}{\beta}} \, U) \supset \beta V$$

whence $SU \supset (\beta - \alpha)V$.

We see thus that Theorem 2.3 is, indeed, a generalization of the subtraction theorem. At the same time the requirement that m be small with respect to q appears as a natural substitute for the condition $\alpha < \beta$.

It would not be difficult to reduce Theorem 2.3 to a simple application of the Induction Theorem. The fact that a considerable simplification of the proof may be achieved by extending the generalization even further seems to speak in favour of striving after a more general if less standard result.

The departure from tradition consists in considering families of relations instead of one fixed relation: it will be set off by the simplicity of the proof.

For the study of relations we shall adopt the same point of view as before: we shall consider approximate relations, i.e. families $S(r)$ of relations depending on a small positive parameter r such that the desired relation appears as a limit of this family in a suitable manner.

Let E and F be two metric spaces and suppose we are given, for each sufficiently small positive r, a relation $S(r) \subset E \times F$. We shall see that a natural notion of a limit of a system of relations is not just the limit of the family $S(\cdot)$ taken as a family of subsets of $E \times F$: a relation smaller in general seems to be suitable.

Given an approximate relation $S(\cdot)$ we define $S(0)$ by the requirement that

$$S(0)^{-1}y = \lim S(r)^{-1}y$$

for each $y \in F$. It follows that the following assertions are equivalent.

1. $xS(0)y$

2. there exists a sequence $x_n \in E$ converging to x and a sequence of positive numbers r_n converging to zero such that $x_n \, S(r_n)y$

3. for every neighbourhood G of x there exist arbitrarily small r such that the set G × {y} intersects S(r).

It is to be noted that S(0) is distinct, in general, from the limit of the family S(r) taken as a family of subsets of E × F; instead, it is only the intersection S(0) ∩ (E × {y}) which may be realized as the limit of the family of intersections S(r) ∩ (E × {y}). Thus, in general, S(0) is a proper subset of the limit lim S(r).

It will be convenient to restate inclusions concerning S in terms of S^{-1} and vice versa. In particular, the following four assertions are equivalent

$$y \in U(Sx,r), \quad x \in S^{-1} U(y,r), \quad Sx \cap U(y,r) \neq \emptyset, \quad d(y,Sx) < r.$$

The geometric significance of these conditions is clear: the set of all pairs [x,y] satisfying the preceding conditions is the graph of a relation which may be obtained from S by allowing imprecisions of order r in measuring in the direction of the y-axis. We shall use the symbol

$$R(S,r)$$

for this modification of the relation S.

In a similar manner we introduce an analogous modification in the direction of the domain space by the following equivalent conditions

$$[x,y] \in D(S,r), \quad x \in U(S^{-1}y,r)$$

$$y \in SU(x,r), \quad S^{-1}y \cap U(x,r) \neq \emptyset, \quad d(x,S^{-1}y) < r.$$

The operations R and D are dual to each other; it is not difficult to verify that

$$D(T,t) = (R(S^{-1},t))^{-1}$$

for every t > 0.

Also, we leave to the reader the proof of the following important fact:

For every relation S the limit of the corresponding approximate relation t → R(S,t) coincides with the closure of S; in particular, a relation S is closed if and only if it coincides with the limit of the approximate relation R(S,·).

Indeed, given $[x,y] \in S^-$ and an arbitrary $r > 0$, there exists a $[u,v] \in S$ for which

$$d(u,x) + d(v,y) < r$$

so that $d(y,Su) \leq d(y,v) < r$ whence

$$u \; R(S,r)y$$

and $d(u,x) < r$. Since r was an arbitrary positive number it follows that $[x,y] \in R(S,0)$.

On the other hand suppose $[x,y] \in R(S,0)$ and let $r > 0$ be given. There exists a $u \in U(x,r)$ and an $s \leq r$ for which $u \; R(S,s)y$ so that $d(y,Su) < s \leq r$. Take a $v \in Su$ for which $d(y,v) < s$. Thus $[u,v] \in S$ and $d(u,x) < r$ and $d(v,y) < r$. Since $r > 0$ was arbitrary it follows that $[x,y] \in S^-$.

Now we are able to state the main result.

2.4   <u>Theorem</u> *Let E and F be two metric spaces. Suppose that, for each $r \in T$, a relation $S(r) \subset E \times F$ is given such that $S(r_1) \subset S(r_2)$ if $r_1 < r_2$. Furthermore let m and q be two positive functions defined on T such that*

$$D(S(m(r)),r) \supset S(q(r))$$

for each $r \in T$.

Suppose further that there exists a rate of convergence w on T such that $m(r) \leq q(w(r))$ for all $r \in T$ and that $m(r) \to 0$ as $r \to 0$. If E is complete then

$$D(S(0), s(r)) \supset S(q(r))$$

for all $r \in T$.

<u>Proof</u> Consider a fixed $r > 0$ and a point $[x_0,y_0] \in S(q(r))$.

For each $t \in T$ set

$$W(t) = S(t)^{-1}y_0$$

and let us prove that

$$W(q(t)) \subset U(W(m(t)),t) \tag{58}$$

45

for all $t \in T$.

If $x \in W(q(t))$ then $[x,y_0] \in S \circ q(t) \subset D(S \circ m(t),t)$ so that

$$S \circ m(t)^{-1} y_0 \cap U(x,t) \neq \emptyset \tag{59}$$

whence $x \in U(W \circ m(t),t)$ and (58) is established.

Now we shall use the monotonicity of the mapping $r \to S(r)$ and the inequality $m(r) \leq q(w(t))$ to prove, for the family $R_1(t) = W(m(t))$, the relation

$$R_1(t) \subset U(R_1(w(t)),w(t)) \tag{60}$$

for all $t \in T$. Using (58), we obtain

$$R_1(t) = W(m(t)) \subset W(q(w(t))) \subset U(W(m(w(t))), w(t)) = U(R_1(w(t)),w(t)).$$

Since $x_0 \in W(q(r))$ we have, by (59)

$$S \circ m(r)^{-1} y_0 \cap U(x_0,r) \neq \emptyset$$

so that there exists an $x_1 \in R_1(r)$ with $d(x_1,x_0) < r$.

It follows from (60) that

$$x_1 \in R_1(r) \subset U(R_1(0), s(w(r)))$$

so that there exists an $x_* \in R_1(0)$ with $d(x_1,x_*) \leq s(w(r))$. Hence $d(x_0,x_*) \leq d(x_0,x_1) + d(x_1,x_*) < s(r)$.

Since $R_1(t) = W(m(t)) = S \circ m(t)^{-1} y_0$ and $m(t) \to 0$ we have $R_1(0) \subset S(0)^{-1} y_0$. The proof is complete. $\qquad \square$

To obtain Theorem 2.3 from Theorem 2.4 we consider a closed relation S and set

$$S(t) = R(S,t).$$

Given $x \in E$, we have

$$D(S(m),r)x = S(m) U(x,r) = U(SU(x,r),m)$$

$$S(t)x = U(Sx,t).$$

Furthermore we have seen already that $S(0) = S^-$ whence, S being closed, we have $S(0) = S$.

3 Multidimensional rates of convergence

Iterative algorithms of type (p,m)

Thus far we have considered iterative algorithms of the form

$$x_{n+1} = Gx_n \quad n = 0,1,2,\ldots$$

There are, however, many important algorithms where the new element x_{n+1} is obtained by means of an explicit expression involving x_n as well as a certain fixed number of the elements preceding x_n. Such an algorithm is described by a mapping $G:X^p \to X$, the recursive formula

$$x_{n+1} = G(x_{n-p+1}, x_{n-p+2},\ldots,x_n), \quad n = 0,1,2,\ldots$$

and a system of p initial starting points which we denote by

$$u_0 = (x_{-p+1}, x_{-p+2},\ldots,x_0), \quad u_0 \in X^p.$$

As an example let us mention the secant method where $p = 2$. An example of a similar procedure with $p = 3$ would be one where the next point is obtained as a solution of a quadratic equation, the parabola being interpolated through three points on the graph of the given function.

It is, sometimes, convenient to simplify some of the formulae by introducing a mapping $\tilde{G}$ of X^p into X^p as follows

$$\tilde{G}(x_1,x_2,\ldots,x_p) = (x_2,x_3,\ldots,x_p, G(x_1,\ldots,x_p)).$$

Starting with u_0, we can consider the iterative algorithm

$$u_{n+1} = \tilde{G}u_n, \quad n = 0,1,2,\ldots \tag{1}$$

and observe that $x_n = P(1)u_n$ for $n = 0,1,2,\ldots$ if $P(1)$ stands for the projection of X^p onto the last coordinate.

We shall also investigate algorithms one step of which consists in the computation of m elements. Such an algorithm is given by a mapping $G:X^p \to X^m$ and an initial vector $u_0 = (x_{-p+1},\ldots,x_0) \in X^p$. The recursion formula assumes different forms according to whether $m \geq p$ or $m < p$.

If $m \geq p$ then the algorithm is given by

$$v_1 = G u_0$$

$$v_{n+1} = G P(p)v_n \quad n = 1,2,\ldots \tag{2}$$

where $P(p):X^m \to X^p$ is the projection on the last p coordinates.

If $m < p$ we set

$$v_1 = G u_0$$

$$u_n = (P(p-m)u_{n-1},v_n), \quad v_{n+1} = Gu_n, \quad n = 1,2\ldots \tag{3}$$

where $P(p-m):X^p \to X^{p-m}$ is the projection on the last $p-m$ components.

Let us show how each of the two cases may be reduced to an iterative algorithm of the form (1) with an appropriate choice of the mapping $\tilde{G}:X^p \to X^p$.

In the case $m \geq p$ we take $\tilde{G} = P(p)G$. Consider the sequence u_n given by (1) and the sequence v_n defined by (2). It is not difficult to see that they are connected by the following formulae

$$u_n = P(p)v_n,$$

$$v_{n+1} = G u_n \quad \text{for } n = 0,1,2,\ldots .$$

In the case $m < p$ we define the mapping $\tilde{G}$ by setting $\tilde{G}u = (P(p-m)u,Gu)$. Then it is easy to see that (1) and (3) generate the same sequence (u_n). This sequence is related to the sequence (v_n) by the formulae $v_n = P(m)u_n = Gu_{n-1}$, $n = 1,2,\ldots$, where $P(m):X^p \to X^m$ is the projection on the last m components.

It is nevertheless possible to give a unified treatment of both cases considered above, a treatment which presents some other advantages as well. Let us begin by introducing a somewhat nonstandard notation: the coordinates of a point $y \in X^{p+m}$ will be denoted by

$$y = (y_{-p+1}, y_{-p+2},\ldots,y_1,y_2,\ldots,y_m).$$

We introduce the projection operators $P_j:X^{p+m} \to X$ defined by the relation

$$P_j y = y_j \quad \text{for } -p + 1 \leq j \leq m.$$

Then the projection on the last p coordinates $P:X^{p+m} \to X^p$ can be written as

$$Py = (P_{m-p+1}y,\ P_{m-p+2}y,\ldots,P_m y)$$

or, $P = (P_{m-p+1},\ P_{m-p+2},\ldots,P_m)$ for short.

With our mapping $G:X^p \to X^m$ we shall associate a mapping $F:X^p \to X^{p+m}$ defined by

$$Fu = (u,\ Gu). \qquad (4)$$

It is not difficult to verify that in both cases discussed above we have $\tilde{G} = PF$.

Let us consider the iterative algorithm

$$z_1 = Fu_0$$

$$z_{n+1} = FPz_n,\ n = 1,2,\ldots\ . \qquad (5)$$

It will be convenient to consider also a $z_0 \in X^{m+p}$ such that $u_0 = Pz_0$.

It is easy to see that if (u_n) is the sequence generated by (1), then, in both cases discussed above, we have

$$u_n = Pz_n,\ n = 0,1,2,\ldots \qquad (6)$$

$$v_n = (P_1 z_n,\ldots,P_m z_n).$$

Until now we have assumed that G (and consequently F) is defined on the whole Cartesian product X^p. In the general case G is defined only on a subset D of X^p. If $u_0 \in D$ then we may obtain the point $z_1 = Fu_0$. If $u_1 = Pz_1$ also belongs to D then we may obtain a new point z_2. We are interested in the case where $Pz_n \in D$ for $n = 1,2,\ldots\ .$

3.1 <u>Definition</u> *Let us consider a mapping* $G:D \subset X^p \to X^m$ *and let* $F:D \subset X^p \to X^{p+m}$ *be the mapping associated with G by* (4). *Denote by* $P:X^{p+m} \to X^p$ *the projection on the last* p *components and define recursively*

$$D_0 = D,\ D_{n+1} = \{u \in D_n;\ PFu \in D_n\},\ n = 0,1,2,\ldots\ .$$

The set $D_\infty = \bigcap_{n \geq 0} D_n$ *is called the set of admissible systems of starting points for the iterative algorithm* (5).

If u_0 belongs to D_∞ then we say that the iterative algorithm (5) *is mean-ingful or well defined.*

In what follows an iterative algorithm of the form (5), where F is of the form (4), will be called an iterative algorithm of type (p,m). In the next paragraph we shall show how the method of nondiscrete induction can be used for the investigation of such iterative algorithms.

Rates of convergence of type (p,m)

We intend to generalize the results of Proposition 1.9 to the case of iterative algorithms of type (p,m). To this effect we shall first define an appropriate notion of a multidimensional rate of convergence. As in Section 1, we shall denote by T either the set of all positive real numbers or a half-open interval of the form $\{t; 0 < t \le t_0\}$. Let w be a function defined on T^p with values in T^m and let $w_j : T^p \to T$, $(j = 1,2,\ldots,m)$ be its components. For all $t \in T^p$ we shall have

$$w(t) = (w_1(t),\ldots,w_m(t)).$$

For the sake of simplicity we shall sometimes write

$$w = (w_1,\ldots,w_m).$$

It will be convenient to denote by w_k, with $1 - p \le k \le 0$ the projection which associates with each vector $t = (t_1,\ldots,t_p) \in T^p$ its (p+k)-th component i.e.

$$w_k(t) = t_{p+k}, \quad 1 - p \le k \le 0.$$

Let us define the iterates $w_j^{(r)}$ of the functions introduced above by setting

$$w_{m+k}^{(0)} = w_k, \quad 1 - p \le k \le 0$$

$$w_j^{(n+1)} = w_j \circ (w_{m-p+1}^{(n)},\ldots,w_m^{(n)}),$$

$$1 - p \le j \le m, \quad n = 0,1,2,\ldots .$$

We shall have evidently $w_j^{(1)} = w_j$ for $1 - p \le j \le m$ and $w_k^{(n+1)} = w_{m+k}^{(n)}$ for $1 - p \le k \le 0$ and $n \ge 0$.

Now let us consider the function

$$\tilde{w} = (w_{m-p+1}, \ldots, w_m) : T^p \to T^p \qquad (6')$$

and let us denote by $\tilde{w}^{(n)}$ its iterates in the sense of usual function composition i.e.

$$\tilde{w}^{(0)}(t) = t, \quad \tilde{w}^{(n+1)}(t) = \tilde{w} \circ \tilde{w}^{(n)}(t), \quad n = 0,1,2,\ldots .$$

We have obviously

$$\tilde{w}^{(n)} = (w_{m-p+1}^{(n)}, \ldots, w_n^{(n)}), \quad n = 0,1,2,\ldots .$$

If we define the iterates $w^{(n)}$ of the mapping w by setting

$$w^{(n+1)} = w \circ \tilde{w}^{(n)}, \quad n = 0,1,2,\ldots$$

then it is easy to see that for all $n \in N$ we have

$$w^{(n)} = (w_1^{(n)}, \ldots, w_m^{(n)}).$$

In Appendix A two examples are given to illustrate how the successive iterates of w are generated.

3.2 **Definition** *A function* $w : T^p \to T^m$ *with the law of iteration described above will be called a rate of convergence of type* (p,m) *on* T, *if the series*

$$s(t) = \sum_{n=1}^{\infty} \sum_{j=0}^{m-1} w_j^{(n)}(t) \qquad (7)$$

is convergent for all $t \in T^p$.

Because $w_0^{(n+1)} = w_m^{(n)}$ for $n = 0,1,2,\ldots$ it follows that for all $t = (t_1, \ldots, t_p) \in T^p$ we can write (7) in the form

$$s(t) = t_p + \sum_{n=1}^{\infty} \sum_{j=1}^{m} w_j^{(n)}(t). \qquad (7')$$

It is easy to see that the function s satisfies the following functional equation

$$s(t) = s(\tilde{w}(t)) + \sum_{j=0}^{m-1} w_j(t) \qquad (8)$$

which is an analogon of the functional equation (1.2). The functional equation (8') turns out to characterize in some sense the rates of convergence of type (p,m). Thus we have the following generalization of Proposition 1.2.

3.3 <u>Proposition</u> *Let* w *be a function defined on* T^p *with values in* T^m *and let us consider the notation introduced at the beginning of this paragraph.*
   Let h *be a function defined on* T^p *with values in* $\mathbb{R}_+$. *Suppose that*

$$h(t) = h(\tilde{w}(t)) + \sum_{j=0}^{m-1} w_j(t) \qquad (8')$$

for all $t \in T^p$.
   Then:

1. w *is a rate of convergence of type* (p,m) *on* T

2. *if the limit* $h(0) = \lim_{\substack{t \to 0 \\ t \in T^p}} h(t)$ *exists then*

$$s(t) = \sum_{n=1}^{\infty} \sum_{j=0}^{m-1} w_j^{(n)}(t) = h(t) - h(0)$$

for all $t \in T^p$.

<u>Proof</u> By iterating the relation (8') we obtain that

$$h(t) = h(\tilde{w}^{(n)}(t)) + \sum_{j=0}^{m-1} w_j(t) + \sum_{j=0}^{m-1} w_j^{(2)}(t) + \ldots + \sum_{j=0}^{m-1} w_j^{(n)}(t)$$

for all $n \in \mathbb{N}$ and $t \in T^p$ and using the fact that the function h has non-negative values we deduce the results stated in the proposition. □

Let w be a rate of convergence of type (p,m) on T. The function $s: T^p \to \mathbb{R}_+$ given by (7) will be called the estimate function corresponding to w. It is convenient to consider also the functions $s_j: T^p \to \mathbb{R}_+$ defined as follows

$$s_0 = s, \quad s_j = s - (w_0 + \ldots + w_{j-1}), \ j = 1, 2, \ldots, m. \qquad (9)$$

With this notation the functional equation (8) becomes

$$s \circ \tilde{w} = s_m. \qquad (8'')$$

Sufficient convergence conditions and error estimates for iterative
algorithms of type (p,m).

Using the notions and the notation introduced in the preceding paragraphs we
shall give a generalization of Proposition 1.9, showing how the method of
nondiscrete induction can be applied to the study of iterative algorithms of
type (p,m).

3.4 **Proposition** *Let p and m be two positive integers. Let X be a complete
metric space and let D be a subset of X^p. Consider a mapping $G:D \to X^m$ and
let $F:D \to X^{p+m}$ be the mapping attached to G by (4). Let us denote by $F_k = P_k F$,
$-p + 1 \leq k \leq m$, the components of F. Let Z be a mapping which associates
with each $t \in T^p$ a subset $Z(t) \subset D$. Let W be a rate of convergence of type
(p,m) on T. Let u_0 be a given element of D and t_0 a given element of T^p.*
  If the following conditions:

$$u_0 \in Z(t_0), \tag{10}$$

$$PFZ(t) \subset Z(\tilde{w}(t)), \tag{11}$$

$$d(F_k u, F_{k+1} u) \leq w_k(t) \tag{12}$$

are satisfied for all $t \in T^p$, $u \in Z(t)$ and $k = 0,1,\ldots,m-1$, then:

1. *u_0 is an admissible system of starting points for the iterative algorithm
    (5);*

2. *there exists a point $x^* \in X$ such that each of the $m + p$ sequences
    $(P_k z_n)_{n \geq 0}$, $-p+1 \leq k \leq m$, converges to x^*;*

3. *the following relations hold for each $n = 1,2,3,\ldots$*

$$Pz_n \in Z(\tilde{w}^{(n)}(t_0)), \tag{13}$$

$$d(P_k z_n, P_{k+1} z_n) \leq w_k^{(n)}(t_0), \quad 0 \leq k \leq m-1, \tag{14}$$

$$d(P_k z_n, x^*) \quad s_k(\tilde{w}^{(n-1)}(t_0)), \quad 0 \leq k \leq m; \tag{15}$$

4. *suppose that for some $n \in \mathbb{N}$ and some $d_n \in T^p$ we have*

$$Pz_{n-1} \in Z(d_n); \tag{16}$$

then

$$d(P_k z_n, x^*) \leq s_k(d_n), \quad 0 \leq k \leq m.$$

(17)

<u>Proof</u> First let us observe that $u_0 \in D_n$ is equivalent to $Pz_n \in D$. Thus statement 1. appears as a consequence of relation (13). Let us prove this relation.
From (10) and (11) it follows that

$$Pz_1 = PFu_0 \in PFZ(t_0) \subset Z(\tilde{w}(t_0)),$$

so that (13) holds for $n = 1$.

Supposing that (13) holds for a certain $n \geq 1$ we have

$$Pz_{n+1} = PFPz_n \in PFZ(\tilde{w}^{(n)}(t_0)) \subset Z(\tilde{w} \circ \tilde{w}^{(n)}(t_0)) = Z(\tilde{w}^{(n+1)}(t_0)).$$

According to (5) and (12) we have

$$d(P_k z_1, P_{k+1} z_1) = d(F_k u_0, F_{k+1} u_0) \leq w_k(t_0).$$

This shows that (14) holds for $n = 1$. If it holds for a certain $n \geq 1$ then, applying (12) for $u = Pz_n$ and $t = \tilde{w}^{(n)}(t_0)$, we obtain

$$d(P_k z_{n+1}, P_{k+1} z_{n+1}) = d(F_k Pz_n, F_{k+1} Pz_n)$$

$$\leq w_k(\tilde{w}^{(n)}(t_0)) = w_k^{(n+1)}(t_0).$$

Thus we have proved that (13) and (14) hold for all $n \in \mathbb{N}$.

To prove statement 2. we note first that (4) and (5) imply that
$P_m z_n = P_0 z_{n+1}$ for all n. Consequently we have

$$d(P_m z_n, P_m z_{n+q}) \leq \sum_{j=1}^{q} \sum_{k=0}^{m-1} d(P_k z_{n+j}, P_{k+1} z_{n+j})$$

$$\leq \sum_{j=1}^{q} \sum_{k=0}^{m-1} w_k^{(n+j)}(t_0).$$

(18)

From the above inequalities it follows that the sequence $(P_m z_n)_{n \geq 1}$ is a Cauchy sequence. Hence there exists a point $x^* \in X$ such that $\lim_{n \to \infty} P_m z_n = x^*$.
From (12) we deduce that $\lim_{n \to \infty} P_k z_n = x^*$ for $-p+1 \leq k \leq m-1$. Letting q to tend to infinity in (18) we infer that the inequalities (15) hold for $k = 0$

and k = m. According to (12) they hold for k = 0,1,2,...,m.

In this manner the first three parts of the proposition are established. Taking n = 1 in (15) it follows that we have proved in particular the following implication:

"if $u_0 \in Z(t_0)$ then $d(P_k Fu_0, x^*) \leq s_k(t_0)$ for k = 0,1,...,m".

To prove 4., suppose n is a natural number for which (16) is satisfied. Replacing in the above implication u_0, t_0 respectively by Pz_{n-1}, d_n we obtain (17). The proof is complete. □

As an immediate consequence of the above proposition we shall obtain a result concerning fixed points of the form

$$\underbrace{(x^*,\ldots,x^*)}_{\text{m times}} = \underbrace{G(x^*,\ldots,x^*)}_{\text{p times}} .$$

This result generalizes the result of Proposition 1.13. There we have supposed that the rate of convergenc was a nondecreasing function. In case of rates of convergence of type (p,m) the monotonicity is taken in the order relations induced in $\mathbb{R}^p$ by $(\mathbb{R}_+)^p$ and in $\mathbb{R}^m$ by $(\mathbb{R}_+)^m$.

3.5 **Proposition** *Let p and m be two positive integers, X a complete metric space, and G a mapping of X^p into X^m . Let $F:X^p \to X^{p+m}$ be the mapping attached to G by (4). Denote $F_k = P_k F$ for $-p+1 \leq k \leq m$ and $F_{m+1} = F_1 PF$. Suppose there exists a nondecreasing rate of convergence of type (p,m), $w:T^p \to T^m$, such that*

$$d(F_k u, F_{k+1} u) \leq w_k(d(F_{-p+1} u, F_{-p+2} u),\ldots,d(F_0 u, F_1 u)) \tag{19}$$

for all $u \in X^p$ and k = 1,2,...,m.

Let $u_0 \in X^p$ be given and set

$$t_0 = (d(F_{-p+1} u_0, F_{-p+2} u_0),\ldots,d(F_0 u_0, F_1 u_0)).$$

Then

(i) *each of the p+m sequences $(P_j z_n)_{n \geq 1}$, $-p+1 \leq j \leq m$, produced by (5) converges to the same point $x^* \in X$ and the following inequalities*

$$d(P_k \, z_n, \, x^*) \leq s_k(d_n) \leq s_k(\tilde{w}^{(n-1)}(t_0))$$

are satisfied for all $k = 0,1,\ldots,m$ *and* $n = 1,2,\ldots,$ *where*

$$d_n = (d(P_{-p+1} \, z_n, \, P_{-p+2} \, z_n),\ldots,d(P_0 \, z_n, \, P_1 \, z_n)).$$

(ii) *if G is continuous at* $(\underbrace{x^*,\ldots,x^*}_{p \text{ times}})$, *then* $(\underbrace{x^*,\ldots,x^*}_{m \text{ times}}) = G(\underbrace{x^*,\ldots,x^*}_{p \text{ times}})$.

<u>Proof</u> Take for every $t \in T^p$

$$Z(t) = \{u \in X^p; \, d(F_k u, F_{k+1} u) \leq w_k(t), \, k = -p+1,\ldots,-1,0\}$$

and apply Proposition 3.4. The inequalities (12) are obviously satisfied (see (19) for $k = 1,2,\ldots,m-1$ and the definition of $Z(t)$ for $k = 0$). The relation (11) follows immediately from (19) observing that $F_k PFu = F_{m+k} u$ for $k = -p+1,\ldots,0,1$. Finally the relation $Pz_{n-1} \in Z(d_n)$ can easily be checked using the fact that $F_k Pz_{n-1} = P_k z_n$ for $k = -p+1,\ldots,0,1$. The proof is complete. □

Iterative procedures of type (p,m) for classes of nonlinear equations

Suppose we are given a nonlinear operator f defined on a subset D_f of a Banach space X with values in a Banach space Y and we want to find a solution of the equation $f(x) = 0$ knowing a system of q approximations of it, $s_0 = (x_{-q+1},\ldots,x_0) \in D_f^q$. Suppose we construct a mapping $G: D \subset D_f^p \to X^m$, with $p \leq q$, such that

$$(\underbrace{x^*,\ldots,x^*}_{m \text{ times}}) = G(\underbrace{x^*,\ldots,x^*}_{p \text{ times}})$$

implies $f(x^*) = 0$. If $u_0 = (x_{-p+1},\ldots,x_0)$ is an admissible system of starting points for the iterative algorithm (5) then we can produce m sequences of points of X converging to a root of the equation $f(x) = 0$. In the following two sections we shall give some examples of how a mapping G with this property can be associated to pairs (f,s_0) belonging to a certain class.

3.6 **Definition** *Let* p,q *and* m *be three positive integers such that* $p \leq q$. *Let C be a class of pairs* (f,s_0) *where f is a nonlinear operator defined on a*

56

subset D_f *of a Banach space* X *with values in a Banach space* Y *and*
$s_0 = (x_{-q+1}, \ldots, x_0)$ *is an element of* D_f^q.

We define an iterative procedure of type (p,m) *for the class* C *as an application which associates with each pair* (f, s_0) *from* C *a mapping*
$G: D \subset D_f^p \to X^m$ *such that:*

1. $u_0 = (x_{-p+1}, \ldots, x_0)$ *is an admissible system of starting points for the iterative algorithm* (5), *where* F *is given by* (4).

2. *each of the* m *sequences* $(P_k z_n)_{n \geq 1}$, $1 \leq k \leq m$ *produced by* (5) *converges to the same point* $x^* \in X$ *so* x^* *is a solution of the equation* $f(x) = 0$.

Having an iterative procedure of type (p,m) for a class C we are interested in finding estimates for the distances $d(P_k z_n, x^*)$ which should be valid for the whole class C. We can do this either by determining m functions $\alpha_k : \mathbb{Z}_+ \to \mathbb{R}_+$, $k = 1, 2, \ldots, m$, such that

$$d(P_k z_n, x^*) \leq \alpha_k(n) \tag{20}$$

for all $(f, s_0) \in C$, $n \in \mathbb{Z}_+$, $k = 1, 2, \ldots, m$, or by finding m functions $\beta_k : \mathbb{R}_+^p \to \mathbb{R}_+$ such that

$$d(P_k z_n, x^*) \leq \beta_k(d(P_{-p+1} z_n, P_{-p+2} z_n), \ldots, d(P_0 z_n, P_1 z_n)) \tag{21}$$

for all $(f, s_0) \in C$, $k = 1, 2, \ldots, m$ and $n = 0, 1, 2, \ldots$.

Let us remark that the right-hand side of (20) can be computed before obtaining the points $P_k z_n$ via the iterative procedure, while the right side of (21) can only be computed after these points have been obtained. This is why the estimates (20) are called apriori estimates and the estimates (21) aposteriori estimates for our iterative procedure of type (p,m) for the class C.

The estimates (20) (or the estimates (21)) will be called sharp if there exists a pair $(f, s_0) \in C$ at which they are attained for all $n \in \mathbb{Z}_+$ and $k = 1, 2, \ldots, m$.

Using Proposition 3.4 we shall obtain, in the following two sections, sharp apriori and aposteriori estimates for several iterative procedures. In proving their sharpness we shall use the following proposition.

3.7 Proposition *Under the hypotheses of Proposition 3.4 suppose that equality is attained in* (15) *for* n = n_0 *and* k = k_0. *Then equality will be attained in* (15) *for all pairs* (n,k) *having the property that either* n > n_0, *or* n = n_0 *and* k ≥ k_0.

Proof Using (14) and the relation $P_m z_n = P_0 z_{n+1}$ we have

$$s_{k_0}(\tilde{w}^{(n_0-1)}(t_0)) = d(P_{k_0} z_{n_0}, x^*)$$

$$\leq \sum_{k=k_0}^{m-1} d(P_k z_{n_0}, P_{k+1} z_{n_0}) + \sum_{n>n_0} \sum_{k=0}^{m-1} d(P_k z_n, P_{k+1} z_n)$$

$$\leq \sum_{k=k_0}^{m-1} w_k^{(n_0)}(t_0) + \sum_{n>n_0} \sum_{k=0}^{m-1} w_k^{(n)}(t_0) = s_{k_0}(\tilde{w}^{(n_0-1)}(t_0)).$$

It follows that $d(P_k z_n, P_{k+1} z_n) = w_k^{(n)}(t_0)$ for all pairs (n,k) having the property that either n > n_0 or n = n_0 and k ≥ k_0. □

4 The secant method and some of its modifications

Description of the algorithms

The secant method, also known under the name of Regula Falsi or the method of chords, is one of the most used iterative procedures for solving non-linear equations. According to A.N. Ostrowski [27], this method is known from the time of early Italian algebraists. In the case of scalar equations, the secant method is better than Newton's method from the point of view of the officiency index (see [28]). The secant method was extended for the solution on nonlinear equations in Banach spaces by A.S. Sergeev [65] and J.W. Schmidt [58], [59]. For this purpose they have used the notion of divided differences of a nonlinear operator (see Appendix B). Later it has been observed that it is possible to use instead of this notion the more general notion of consistent approximation of the derivative (see [6] and [60]).

Let f be a nonlinear operator between two Banach spaces and let δf be a consistent approximation of its derivative (see Appendix B for the definition of the consistent approximation). Then the secant method can be described by the recurrence scheme

$$x_{n+1} = x_n - \delta f(x_{n-1}, x_n)^{-1} f(x_n), \quad n = 0,1,2,\ldots \tag{1}$$

Under certain assumptions on the operator f and the initial points x_0 and x_{-1} it is possible to prove that the sequence $(x_n)_{n \geq 0}$ given by (1) converges to a root x^* of the equation $f(x) = 0$, the R-order of convergence being $(1 + \sqrt{5})/2$. The iterative algorithm (1) requires the inversion of a linear operator at each step. If this algorithm is replaced by

$$x_{n+1} = x_n - \delta f(x_{-1}, x_0)^{-1} f(x_n), \quad n = 0,1,2,\ldots \tag{2}$$

there is just one inversion to be performed but the convergence is only linear. A procedure intermediate between these two consists in fixing a natural number m and keeping the same linear operator for sections of the process consisting of m steps each. It may be described as follows:

Given two points $x_0^m = x_0$ *and* $x_0^{m-1} = x_{-1}$, *construct* m *sequences* $(x_n^j)_{n \geq 1}$, $1 \leq j \leq m$, *by the algorithm:*

$$x_n^0 = x_{n-1}^m$$

$$x_n^{j+1} = x_n^j - \delta f(x_{n-1}^{m-1}, x_{n-1}^m)^{-1} f(x_n^j) \tag{3}$$

$$j = 0,1,\ldots,m-1$$

$$n = 1,2,\ldots .$$

J.W. Schmidt and H. Schwetlick [61] made a local analysis for this iterative procedure in the case where $\delta f(x,y)$ was a divided difference of f on x and y. They have shown that the order of convergence of the procedure is $\frac{1}{2}(m + (m^2 + 4)^{\frac{1}{2}})$. The parameter m can be chosen according to the dimension of the space in order to maximize the numerical efficiency.

The class $C(h_0, q_0, r_0)$.

We shall study the iterative algorithms (1), (2) and (3) for triplets (f, x_0, x_1) belonging to the class $C(h_0, q_0, r_0)$ defined as follows:

4.1 <u>Definition</u> *Let* $h_0 > 0$, $q_0 \geq 0$ *and* $r_0 \geq 0$ *be three given real numbers. We say that a triplet* (f, x_0, x_{-1}) *belongs to the class* $C(h_0, q_0, r_0)$ *if:*

(c_1) f *is a nonlinear operator defined on a subset* D_f *of a Banach space* X *with values in a Banach space* Y.

(c_2) x_0 *and* x_{-1} *are two points of* D_f *such that*

$$|x_0 - x_{-1}| \leq q_0 < \mu \tag{4}$$

(c_3) f *is Fréchet differentiable in the open ball* $U := U(x_0, \mu)$ *and continuous on its closure* $\bar{U}$.

(c_4) *There exists a mapping* $\delta f: U \times U \to B(X,Y)$ *such that* $S_0 := \delta f(x_{-1}, x_0)$ *is invertible,*

$$T_0 := \delta f(x_{-1}, x_0)^{-1} \in B(Y,X) \quad and$$

$$|T_0(\delta f(x,y) - f'(z))| \leq h_0(|x-z| + |y-z|) \text{ for all } x,y,z \in U. \tag{5}$$

(c_5) *The following inequalities are satisfied:*

$$|T_0 f(x_0)| \leq r_0,\tag{6}$$

$$h_0 q_0 + 2\sqrt{h_0 r_0} \leq 1,\tag{7}$$

$$\mu \geq \frac{1}{2h_0}\left(1 - h_0 q_0 - \sqrt{(1 - h_0 q_0)^2 - 4h_0 r_0}\right) =: \mu_0\tag{8}$$

We shall show that if $(f, x_0, x_{-1}) \in C(h_0, q_0, r_0)$ then the equation $f(x) = 0$ has a solution x^* which is unique in a certain neighbourhood of x_0. We shall also show that all the sequences produced by (1), (2) and (3) converge to this solution. Before proving these results we intend to make some remarks on the conditions defining the class $C(h_0, q_0, r_0)$. Condition (5) means that $T_0 \delta f$ is a consistent approximation of $T_0 f$. From (5) it follows that

$$|T_0(f(x) - f(y) - \delta f(u,v)(x - y))|$$

$$\leq h_0(|x - v| + |y - v| + |u - v|)\,|x - y|\tag{9}$$

for all $x, y, u, v \in U$ (see Appendix B). This inequality will be often used in the sequel.

The constant h_0 appearing in (5) depends on μ in general. In (8) we require μ to be greater than μ_0 which depends on h_0. It is then useful to note that $\mu_0 < r_0 + \sqrt{r_0(q_0 + r_0)}$ for all $h_0 > 0$, so that we could take $\mu = r_0 + \sqrt{r_0(q_0 + r_0)}$. The most restrictive condition from the definition of the class $C(h_0, q_0, r_0)$ seems to be inequality (7). This inequality is satisfied only if q_0 and r_0 are small enough. In practical applications q_0 can be taken as small as wanted, because, having an initial point x_0 we can choose x_{-1} very close to it, but r_0 can be taken small only if the initial approximation is "good enough" (see (6)). It is not so easy to find such an initial point! However it turns out that condition (7) is optimal in some sense. Indeed, we shall show that if this condition is not satisfied then it is no longer possible to ensure the existence of a root of the equation $f(x) = 0$.

The simplified secant method

The iterative algorithm (2) is often called the simplified secant method. It was first studied by S. Ulm [68]. In what follows we shall prove that the simplified secant method represents an iterative procedure of type (1.1) for the class $C(h_0,q_0,r_0)$ and shall find sharp apriori and aposteriori error estimates for it (see Section 3). First of all let us associate with this iterative algorithm a rate of convergence of type (1.1).

4.2 <u>Lemma</u> *If* $a \geq 0$ *and* $\beta > 2a$ *then the function*

$$w(r) = r(1 + \frac{r}{\beta} - \frac{2}{\beta} (a^2 + \beta r)^{\frac{1}{2}}) \tag{10}$$

is a rate of convergence of type (1.1) on the interval $(0,r_0)$ *where* $r_0 = \beta-2a$, *and the corresponding estimate function is given by*

$$s(r) = (a^2 + \beta r)^{\frac{1}{2}} - a \tag{11}$$

In particular, if $h_0 > 0$, $q_0 \geq 0$, $r_0 \geq 0$ *are three numbers satisfying condition (7), then* $\beta = h_0^{-1}$ *and*

$$a = \frac{1}{2h_0} ((1 - h_0 q_0)^2 - 4h_0 r_0)^{\frac{1}{2}} \tag{12}$$

is a possible choice.

<u>Proof</u> Apply Proposition 1.2. □

Using the above lemma and Proposition 1.9 we obtain the following result

4.3 <u>Theorem</u> *Let* w *and* s *be the functions defined in Lemma 4.2. If* $(f,x_0,x_{-1}) \in C(h_0,q_0,r_0)$ *then the iterative algorithm (2) yields a sequence* $(x_n)_{n \geq 0}$ *of points belonging to the open sphere* $U(x_0,\mu_0)$, *which converges to a root* x^* *of the equation* $f(x) = 0$ *and the following estimates hold:*

$$|x_n - x^*| \leq s(w^{(n)}(r_0)), \quad n = 0,1,2,\ldots \tag{13}$$

$$|x_n - x^*| \leq s(|x_n - x_{n-1}|) - |x_n - x_{n-1}|, \quad n = 1,2,3,\ldots \ . \tag{14}$$

<u>Proof</u> Let us consider the mappings $G:U(x_0,\mu_0) \to X$ and $Z:(0,r_0] \to \exp X$ given by the following relations:

$$Gx = x-T_0 f(x), \quad Z(r) = \{x \in X; \ |x-x_0| \leq s(r_0)-s(r), |T_0 f(x)| \leq r\}. \tag{15}$$

Observing that $s(r_0) = \mu_0$, it follows that $Z(r) \subset U(x_0, \mu_0)$. If $r \in (0, r_0]$, $x \in Z(r)$ and $x' = Gx$, then we have

$$|x' - x_0| \leq |x' - x| + |x - x_0| \leq r + s(r_0) - s(r)$$

$$= s(r_0) - s(w(r)).$$

The relation $x' = Gx$ is equivalent to $f(x) + S_0(x' - x) = 0$, so that, using (9), we obtain

$$|T_0 f(x')| = |T_0(f(x') - f(x) - S_0(x' - x))|$$

$$\leq h_0(|x' - x_0| + |x - x_0| + |x_0 - x_{-1}|)\, |x' - x|$$

$$\leq h_0(2s(r_0) - 2s(r) + r + q_0)r = w(r).$$

From the above relations it follows that the hypotheses (i) and (ii) of Proposition 1.9 are satisfied. Thus the sequence $(x_n)_{n \geq 0}$ converges to a point $x^* \in X$. The estimates (13) follow then from (1.10), while, corresponding to (1.8) and (1.9), we have

$$x_{n-1} \in Z(w^{(n-1)}(r_0)), \quad |x_n - x_{n-1}| \leq w^{(n-1)}(r_0), \quad n = 1,2,3,\ldots . \quad (16)$$

Using the fact that s is increasing on $(0, r_0]$ we deduce from the above relations that

$$x_{n-1} \in Z(|x_n - x_{n-1}|),$$

so that according to point 3. of Proposition 1.9 it follows that the aposteriori estimates (14) are true for $n = 1,2,\ldots$.

To finish the proof of the theorem we observe that, by letting n tend to infinity in (2), we obtain $f(x^*) = 0$. □

From the above theorem it follows that the mapping which associates with each triplet

$$(f, x_0, x_{-1}) \in C(h_0, q_0, r_0)$$

the operator G given by (15) is an iterative procedure of type (1.1) for the class $C(h_0, q_0, r_0)$ in the sense of Definition 3.6. It also follows that the functions

$$\alpha(n) = s(w^{(n)}(r_0)) \text{ and } \beta(r) = -r - s + \sqrt{a^2 + h_0^{-1}r}$$

are respectively apriori and aposteriori error estimates for this iterative procedure. In the following proposition we shall prove that these error estimates are sharp.

4.4 <u>Proposition</u> *If* $h_0 > 0$, $q_0 \geq 0$, $r_0 \geq 0$ *are three constants satisfying inequality (7), then there exist a function* $f:\mathbb{R} \to \mathbb{R}$ *and two points* $x_0, x_{-1} \in \mathbb{R}$ *such that the triplet* (f, x_0, x_{-1}) *belongs to* $C(h_0, q_0, r_0)$ *and for this triplet the estimates (13) and (14) are attained for all* n.

<u>Proof</u> Take $f(x) = x^2 - a^2$, where a is given by (12),

$$x_0 = \frac{1 - h_0 q_0}{2h_0} \,, \quad x_{-1} = \frac{1 - h_0 q_0}{2h_0}$$

$$\mu = r_0 + \sqrt{r_0(q_0 + r_0)}, \quad \delta f(x,y) = (f(x) - f(y))/(x-y).$$

The fact that $(f, x_0, x_{-1}) \in C(h_0, q_0, r_0)$ can easily be checked. Let $(x_n)_{n \geq 0}$ be the sequence associated with (f, x_0, x_{-1}) by the iterative algorithm (2) - this sequence obviously converges to a. We have

$$x_0 - a = \omega_0 = s(r_0).$$

This means that the estimate (13) is attained for n = 0. According to Proposition 1.11 this estimate will be attained for all n ≥ 0. The function $s(r) = -a + \sqrt{a^2 + h_0^{-1}r}$ is increasing on the interval $(0, r_0]$ so that by virtue of Proposition 1.10 the error estimates (14) will be also attained for all n ≥ 1.

Existence and uniqueness of the solution

In this paragraph we shall prove a theorem concerning the uniqueness of the solution of the equation f(x) = 0 for triplets (f, x_0, x_{-1}) belonging to the class $C(h_0, q_0, r_0)$. First of all we shall prove a lemma containing some information about the behaviour of the sequence $(s(w^{(n)}(r_0)))_{n \geq 0}$ which appears in (13). The cases a > 0 and a = 0 are considered separately. Let us observe that a = 0 if and only if we have equality in (7).

4.5 <u>Lemma</u> *Suppose the hypotheses of Lemma 4.2 are verified.*

(i) *If a > 0, then the inequalities*

$$\frac{2r_0}{1-h_0q_0+2h_0a}\,[h_0(q_0 + r_0)]^n \leq s(w^{(n)}(r_0))$$

$$\leq \frac{r_0}{2h_0a}\,(1 - 2h_0a)^n \qquad (17)$$

hold for n = 0,1,2,...

(ii) *If a = 0 then the following estimates are satisfied for every* n=1,2,3...

$$\frac{1}{n+1}\sqrt{\frac{r_0}{h_0}} \leq s(w^{(n)}(r_0)) \leq \frac{1}{n+2}\,\frac{1}{h_0} \qquad (18)$$

<u>Proof</u> (i) Observing that for r ∈ (0,r_0] we have

$$h_0(q_0 + r_0)r \leq w(r) \leq (1 - 2h_0a)r$$

and, using the fact that w is increasing on (0,r_0], it is possible to show that

$$r_0[h_0(q_0 + r_0)]^n \leq w^{(n)}(r_0) \leq r_0(1 - 2h_0a)^n, \; n = 0,1,2,... \;.$$

The estimates (17) follow then as a consequence of the inequalities

$$\frac{2r}{1-h_0q_0+2h_0a} \leq s(r) \leq \frac{r}{2ah_0}\,,$$

which are valid for every r ∈ (0,r_0].

(ii) The inequalities (18) can easily be proved by induction observing that for a = 0 we have $h_0[s(r)]^2 = r$ and consequently

$$s(w^{(n+1)}(r_0)) = s(w^{(n)}(r_0)) - h_0[s(w^{(n)}(r_0))]^2$$

for all n ≥ 0. □

Using Theorem 4.3 and the above lemma we obtain the following result about the existence and uniqueness of the solution for nonlinear equations.

4.6 Theorem *If the triplet (f, x_0, x_{-1}) belongs to the class $C(h_0, q_0, r_0)$ then the equation $f(x) = 0$ has a root x^* belonging to the closed ball*

$$W = \{x \in X; \; |x - x_0| \leq \omega_0\}$$

and this root is unique in the set $V := \bar{U} \cap U(x_0, \mu_0 + 2a)$ if $a > 0$, or in W if $a = 0$.

Proof The existence has been proved in Theorem 4.3. Suppose $a > 0$ and let $x^* \in \bar{U}$ and $y^* \in V$ be two solutions of the equation $f(x) = 0$. Let us denote

$$A_* = \int_0^1 f'(y^* + t(x^* - y^*))dt.$$

According to (5) we have

$$|I - T_0 A_*| = |T_0(S_0 - A_*)|$$

$$\leq h_0(|y^* - x_0| + |x^* - x_0| + |x_0 - x_{-1}|)$$

$$< h_0(2\mu_0 + 2a + q_0) = 1.$$

It follows that A_* is invertible. On the other hand

$$A_*(x^* - y^*) = f(x^*) - f(y^*) = 0$$

(see Appendix B). Hence $x^* = y^*$.

Now let us suppose that $a = 0$. Consider the sequence $(x_n)_{n \geq 0}$ generated by the iterative algorithm (2). From Theorem 4.3 it follows that this sequence converges to a root $x^* \in W$ of the equation $f(x) = 0$. It also follows that $|x_n - x_0| \leq s(r_0) - s(w^{(n)}(r_0))$, for $n = 0, 1, 2, \ldots$ (see (16)). In our case

$$s(r_0) = \mu_0 = \sqrt{h_0^{-1} r_0}.$$

If $y^* \in W$ is another solution of the equation $f(x) = 0$ then we can write

$$|x_{n+1} - y^*| = |T_0(f(x_n) - f(y^*) - S_0(x_n - y^*))|$$

$$\leq h_0(|x_n - x_0| + |y^* - x_0| + |x_{-1} - x_0|) \, |x_n - y^*|$$

$$\leq h_0(2\mu_0 + q_0 - s(w^{(n)}(r_0)))|y^* - x_n|$$

$$\leq (1 - h_0 s(w^{(n)}(r_0)))\, |y^* - x_n|$$

$$\leq |x_1 - y^*| \prod_{j=1}^{n} (1 - h_0 s(w^{(j)}(r_0))).$$

From (18) it follows that $\lim\limits_{n\to\infty} \prod_{j=1}^{n} (1 - h_0 s(w^{(j)}(r_0))) = 0$. Thus we infer that $y^* = \lim\limits_{n\to\infty} x_n = x^*$. $\qquad\square$

In the following proposition we shall show that the results concerning the domain of uniqueness obtained above are sharp within the class $C(h_0, q_0, r_0)$.

4.7 **Proposition** *Let $h_0 > 0$, $q_0 \geq 0$, $r_0 \geq 0$ be three numbers satisfying condition (7) and let a be the constant defined by (12).*

(i) *If $a > 0$ then there exist a function $f: \mathbb{R} \to \mathbb{R}$ and four points x_0, x_{-1}, x^*, $y^* \in \mathbb{R}$ such that $(f, x_0, x_{-1}) \in C(h_0, q_0, r_0)$,*

$$f(x^*) = f(y^*) = 0, \quad |x_0 - x^*| = \mu_0, \quad |x_0 - y^*| = \mu_0 + 2a.$$

(ii) *If $a = 0$, then for each $\varepsilon > 0$ there exist a function $f_\varepsilon: \mathbb{R} \to \mathbb{R}$ and four points $x_0, x_{-1}, x^*, y^*_\varepsilon \in \mathbb{R}$ such that*

$$(f_\varepsilon, x_0, x_{-1}) \in C(h_0, q_0, r_0), \quad f_\varepsilon(x^*) = f_\varepsilon(y^*_\varepsilon) = 0,$$

$$|x^* - x_0| = \mu_0, \quad |y^*_\varepsilon - x_0| = \mu_0 + \varepsilon.$$

Proof (i) Take $f(x) = x^2 - a^2$, $x_0 = \dfrac{1 - h_0 q_0}{2h_0}$,

$$x_{-1} = \frac{1 + h_0 q_0}{2h_0}, \quad x^* = a, \quad y^* = -a.$$

(ii) Take

$$f_\varepsilon(x) = \begin{cases} x^2, & \text{if } x \geq \dfrac{-\varepsilon}{2 + \sqrt{2}} \\[2ex] -x^2 - \dfrac{4\varepsilon}{2 + \sqrt{2}}\, x - 2\left(\dfrac{\varepsilon}{2 + \sqrt{2}}\right)^2, & \text{if } x < \dfrac{-\varepsilon}{2 + \sqrt{2}}, \end{cases}$$

$$x_0 = \frac{1 - h_0 q_0}{2h_0}, \quad x_{-1} = \frac{1 + h_0 q_0}{2h_0}, \quad x^* = 0, \quad y_\varepsilon^* = -\varepsilon. \qquad \square$$

We have already remarked the special position of inequality (7) among the conditions defining the class $C(h_0, q_0, r_0)$. In what follows we shall show that this condition is necessary (in a certain sense) for the existence of the solution.

4.8 **Proposition** *Let $h_0 > 0$, $q_0 \geq 0$, $r_0 > 0$ be three numbers which do not satisfy condition (7). Then there exists a function $f: \mathbb{R} \to \mathbb{R}$ and two points x_0, $x_{-1} \in \mathbb{R}$ such that:*

(i) *conditions (c_1) - (c_4) as well as inequality (6) are satisfied;*

(ii) *the equation $f(x) = 0$ has no solution.*

Proof First of all we observe that the inequality $h_0 q_0 + 2\sqrt{h_0 r_0} > 1$ is equivalent to $h_0^{-1} < q_0 + 2r_0 + 2\sqrt{r_0(q_0 + r_0)}$. We consider two cases:

If $q_0 + 2r_0 - 2\sqrt{r_0(q_0 + r_0)} < h_0^{-1} < q_0 + 2r_0 + 2\sqrt{r_0(q_0 + r_0)}$, then we take

$$f(x) = h_0 x^2 + \frac{1}{4h_0}(2h_0(q_0 + 2r_0) - 1 - h_0^2 q_0^2),$$

$$x_0 = \frac{1 - h_0 q_0}{2h_0}, \quad x_{-1} = \frac{1 + h_0 q_0}{2h_0}$$

and if $0 < h_0^{-1} \leq q_0 + 2r_0 - 2\sqrt{r_0(q_0 + r_0)}$, then we shall take

$$f(x) = q_0^{-1} x^2 + r_0, \quad x_0 = 0, \quad x_{-1} = q_0. \qquad \square$$

The secant method

In studying the secant method we shall use the rate of convergence of type (2.1) given in the following lemma.

4.9 **Lemma** *Let T be the whole positive axis and let a be a non-negative number. Then the function*

$$w(q,r) = \frac{r(q + r)}{r + 2(r(r+q) + a^2)^{\frac{1}{2}}} \qquad (19)$$

68

is a rate of convergence of type (2.1) on T *and the corresponding estimate
function is given by*

$$s(q,r) = r - a + \sqrt{r(q + r) + a^2}. \tag{20}$$

Proof It is easy to check that $s(q,r) = s(r,w(q,r)) + r$ for all $q, r \in T$.
The rest follows from Proposition 3.3.   □

At each step of the iterative algorithm (1) we have to invert a linear
operator. In proving the invertibility of these operators we shall use the
following well known result:

4.10 Lemma   *Let* X *and* Y *be two Banach spaces and let us denote by* $B(X,Y)$
the Banach space of all bounded linear operators from X *into* Y. *If*
$L_0 \in B(X,Y)$ *is invertible and if* $L \in B(X,Y)$ *satisfies* $|L| < |L_0^{-1}|^{-1}$ *then the
operator* L_0-L *is invertible as well and*

$$|(L_0 - L)^{-1}| \leq (1 - |L| \cdot |L_0^{-1}|)^{-1} |L_0^{-1}|. \tag{21}$$

Now we can state our result concerning the convergence of the secant method.

4.11 Theorem *Let* $h_0 > 0$, $q_0 \geq 0$, $r_0 \geq 0$ *be three numbers verifying condition (7). Let* a *be the constant given in (12) and consider the functions*
w *and* s *obtained in Lemma 4.9.*

If $(f,x_0,x_{-1}) \in C(h_0,q_0,r_0)$, *then the iterative procedure (1) yields a
sequence* $(x_n)_{n \geq 0}$ *of points from the open ball* $U(x_0,\mu_0)$ *which converges to a
root* x^* *of the equation* $f(x) = 0$ *and the following estimates hold:*

$$|x_n - x^*| \leq s(\tilde{w}^{(n)} (q_0,r_0)), \; n = 0,1,2,\ldots \tag{22}$$

$$|x_n - x^*| \leq [a^2 + |x_n - x_{n-1}| \, (|x_{n-1} - x_{n-2}|$$

$$+ |x_n - x_{n-1}|)]^{1/2} - a, \; n = 1,2,\ldots \tag{23}$$

where, in accordance with the notation introduced in Section 3, $\tilde{w}(q,r) =
(r,w(q,r))$.

Proof We shall use Proposition 3.4 for the case $p = 2$ and $m = 1$. The

69

obvious choice for D and G is

$$D = \{u = (y,x) \in U^2; \delta f(y,x) \text{ is invertible}\}$$

$$Gu = x - \delta f(y,x)^{-1} f(x), \quad u = (y,x) \in U^2. \tag{24}$$

As in Lemma 4.9 we denote by T the whole positive axis. For every $t = (q,r) \in T^2$ we consider the set

$$Z(t) = \{u = (y,x) \in X^2; y \in U, |x - y| \le q,$$

$|x - x_0| \le \mu_0 - s(t)$, the linear operator $S = \delta f(y,x)$ is invertible and $|S^{-1} f(x)| \le r\}$.

Set $t_0 = (q_0, r_0)$. From (12) and (20) it follows that $s(t_0) = \mu_0$. The function s being increasing we have

$$0 < s(t) < \mu_0 \text{ for all } 0 < t < t_0.$$

It is easy to see that $Z(t) \subset D$ and $u_0 = (x_{-1}, x_0) \in Z(t_0)$. Let us prove now that if $u = (y,x) \in Z(t)$ then $(x, Gu) \in Z(w(t))$. To this end set $z = Gu$ and let us prove the following facts:

$$x \in U, |z - x| \le r, \tag{25}$$

$$|z - x_0| \le \mu_0 - s(\tilde{w}(t)), \tag{26}$$

The operator $S_1 = \delta f(x,z)$ is invertible and $|S_1^{-1} f(z)| \le w(t)$. (27)

The relations (25) are immediate consequences of the fact that $u \in Z(t)$, because $z - x = -S^{-1} f(x)$. Using the relation $s(\tilde{w}(t)) = s(t) - r$ we may write

$$|z - x_0| \le |z - x| + |x - x_0| \le r + \mu_0 - s(t) = \mu_0 - s(\tilde{w}(t)).$$

In order to prove (27), let us note first that according to (5) we have:

$$|T_0(S_0 - S_1)| \le |T_0(S_0 - f'(x))| + |T_0(f'(x) - S_1)|$$

$$\le h_0(|x - x_0| + |x - x_{-1}| + |x - z|)$$

$$\leq h_0(q_0 + r + 2(\mu_0 - s(t)) = 1 - h_0(r + 2\sqrt{r(q + r) + a^2}).$$

Applying now Lemma 4.10 it follows that the linear operator S_1 is invertible and

$$|(T_0S_1)^{-1}| \leq \frac{1}{h_0(r + 2\sqrt{r(q + r) + a^2})}. \tag{28}$$

On the other hand, (24) implies the identity

$$f(z) = f(z) - f(x) - \delta f(y,x) (z - x)$$

and using (9) we obtain

$$|T_0 f(z)| \leq h_0(|z - x| + |x - y|) \, |z - x| \leq h_0 r(q + r). \tag{29}$$

Finally from (28) and (29) we have

$$|S_1^{-1}f(z)| = |(T_0S_1)^{-1}T_0f(z)| \leq w(t).$$

Thus we have checked the validity of (25), (26) and (27). It follows that the hypotheses of Proposition 3.4 are satisfied in our case. Consequently, the sequence $(x_n)_{n \geq 0}$ produced by (1) converges to a point $x^* \in \bar{U}(x_0, \mu)$ and the apriori estimates (22) are satisfied. Moreover we have:

$$(x_{n-2}, x_{n-1}) \in Z(\tilde{w}^{(n-1)} (t_0)), \quad n = 1,2,3,\ldots$$

$$|x_{k+1} - x_k| \leq w^{(k)}(t_0), \quad k = 0,1,2,\ldots .$$

The function s being increasing it is easy to infer from the above relation that

$$(x_{n-2}, x_{n-1}) \in Z(|x_{n-1} - x_{n-2}|, |x_n - x_{n-1}|), \quad n = 1,2,3,\ldots .$$

According to point 4. of Proposition 3.4 the above relation implies the inequality (23).

We still have to prove that x^* is the solution of the equation $f(x) = 0$. This follows easily if we substitute in (29) $t = \tilde{w}^{(n)}(t_0)$, $z = x_{n+1}$ and let n tend to infinity. $\qquad \square$

The above theorem shows that the mapping which associates with each triplet (f,x_0,x_{-1}) from $C(h_0,q_0,r_0)$ the operator G given by (24) is an iterative procedure of type (2.1) for the class $C(h_0,q_0,r_0)$. The inequalities (22) and (23) represent error estimates for this iterative procedure. It is easy to prove that the aposteriori error estimates (23) are better than the apriori error estimates (22). Also, it is possible to prove that these estimates are exact for the triplet (f,x_0,x_{-1}) considered in Proposition 4.4. In the terminology introduced in Section 3 this means that our estimates are sharp error estimates for the secant method in the class $C(h_0,q_0,r_0)$.

An iterative procedure of type (2.m)

In what follows we shall study the iterative algorithm (3). This can be considered as an iterative algorithm of type (2.m). Indeed, for a given triplet $(f,x_0,x_{-1}) \in C(h_0,q_0,r_0)$ and for $u = (y,x) \in U^2$ we may define recursively

$$F_{-1}u = y, \quad F_0u = x; \quad F_{j+1}u = F_ju - \delta f(y,x)^{-1}f(F_ju), \quad j = 0,1,\ldots,m-1.$$
$$(30)$$

Let D be the set of those $u \in U^2$ for which the above formula makes sense (i.e. $\delta f(y,x)$ is invertible and $F_ju \in U$ for $j = 0,1,\ldots,m-1$). The mapping $F:D \to X^{m+2}$ given by $Fu = (F_{-1}u, F_0u,\ldots,F_mu)$ is clearly of the form (3.4). It is easy to see that the sequences obtained via the iterative algorithms (3) and (3.5) are connected by the relation

$$x_n^j = P_jz_n \text{ for } j = 0,1,\ldots,m \text{ and } n = 1,2,3,\ldots .$$

We shall study the iterative algorithm (3) with the help of Proposition 3.4. First we shall associate with this algorithm a rate of convergence of type (2.m).

There are some differences between the cases $m = 1$ and $m \geq 2$, but we can study them together if we make the following convention: if an algorithm requires, at a certain stage, the computation of a quantity Q_k for $k = 0,1,\ldots,p$, and if p happens to be negative, ignore this instruction and pass to the next one; in the same sense the sum $a_0 + a_1 + \ldots + a_p$ will be taken equal to zero if p is negative.

For $m = 1$ the iterative algorithm (3) reduces to the secant method (1). Thus, the results to be presented in the sequel generalize the results obtained in the preceding paragraph.

4.12 Lemma *Let* T *denote the set of all positive real numbers, let* a *be a non-negative real number, and let* m *be a positive integer. For all* $t = (q,r) \in T^2$ *consider the functions*

$$\phi(t) = r + \sqrt{r(q + r) + a^2}, \quad w_{-1}(t) = q, \quad w_0(t) = r$$

and define recursively

$$w_{k+1}(t) = \frac{w_k(t)(w_{-1}(t) + w_k(t) + 2(w_0(t) + \cdots + w_{k-1}(t)))}{2\phi(t) + w_{-1}(t)}$$

$$w_m(t) = \frac{w_{m-1}(t)(w_{-1}(t) + w_{m-1}(t) + 2(w_0(t) + \cdots + w_{m-2}(t)))}{2\phi(t) - 2(w_0(t) + \cdots + w_{m-2}(t)) - w_{m-1}(t)}$$

Then the mapping $w(t) = (w_1(t),\ldots,w_m(t))$ *is a rate of convergence of type* $(2.m)$ *on* T *and the corresponding estimate function is given by*

$$s(t) = \phi(t) - a.$$

Proof Prove by induction that

$$w_k(t)(w_{-1}(t) + w_k(t) + 2(w_0(t) + \cdots + w_{k-1}(t)))$$

$$= (\phi(t) - \sum_{j=0}^{k} w_j(t))^2 - a^2$$

for $k = 0,1,\ldots,m-1$ and then apply Proposition 3.3. □

4.13 Theorem *Consider the rate of convergence obtained in Lemma 4.12 with the constant* a *given by (12), and the mappings attached to it by (3.6) and (3.9). If* $(f,x_0,x_{-1}) \in C(h_0,q_0,r_0)$ *then the iterative procedure (3), with starting points* $x_0^m = x_0$, $x_0^{m-1} = x_{-1}$, *produces* $m+1$ *sequences* $(x_n^j)_{n\geq 0}$, $0 \leq j \leq m$, *which converge to a root* x^* *of the equation* $f(x) = 0$ *and the following inequalities*

$$|x_n^j - x^*| \leq s_j(\tilde{w}^{(n-1)}(q_0,r_0)), \tag{31}$$

$$|x_n^j - x^*| \le s_j(|x_{n-1}^{m-1} - x_{n-1}^m|, |x_n^0 - x_n^1|), \tag{32}$$

hold for all $j = 0,1,\ldots,m$ *and* $n = 1,2,3,\ldots$.

<u>Proof</u> Denote $t_0 = (q_0, r_0)$ and assign to each $t = (q,r) \in T^2$ a subset of X^2 defined by

$$Z(t) = \{u = (y,x) \in X^2; \; y \in U, \; |y - x| \le q, \tag{33}$$

$$|x - x_0| \le s(t_0) - s(t),$$

$$S := \delta f(y,x) \text{ is invertible,}$$

$$|S^{-1}f(x)| \le r, \; |(T_0 S)^{-1}| \le \frac{1}{h_0(2\phi(t) + q}\}.$$

Because $s(t_0) = \mu_0$ it follows that $Z(t) \subset U^2$ for all t. Our theorem will be proved if we show that $Z(t) \subset D$ and that conditions (3.10), (3.11), (3.12) and (3.16) (with $d_n = (|x_{n-1}^{m-1} - x_{n-1}^m|, |x_n^0 - x_n^1|)$) are satisfied. First of all, if u_0 stands for (x_{-1}, x_0) we clearly have $u_0 \in Z(t_0)$. Let us prove now that $u \in Z(t)$ implies:

$$F_k u \in U \text{ for } -1 \le k \le m \tag{34}$$
and
$$|F_k u - F_{k+1} u| \le w_k(t), \text{ for } -1 \le k \le m-1. \tag{35}$$

For $k = -1$ these relations reduce to $y \in U$ and $|y - x| \le q$; for $k = 0$ they follow from $x \in U$ and $|\delta f(y,x)^{-1} f(x)| \le r$.

Consider now an i, $0 \le i \le m-1$, and suppose that (34) and (35) hold for $k = -1, 0, \ldots, i$. We have

$$|x_0 - F_{i+1} u| \le |x_0 - x| + |F_{i+1} u - x|$$

$$\le |x - x_0| + \sum_{j=0}^{i} |F_{j+1} u - F_j u|$$

$$\le s(t_0) - s(t) + \sum_{j=0}^{i} w_j(t) = s(t_0) - s_{i+1}(t),$$

so that $F_{i+1} u \in U$ as well; this establishes (34). Let us remark that from (33) and (34) it follows that $Z(t) \subset D$. To simplify the formulae set

$S_m = \delta f(F_{m-1}u, F_m u)$, $f_j = f(F_j u)$. The relation defining $F_{i+1}u$ may be re-written in the form $f_i = S(F_i u - F_{i+1}u)$. Using this relation together with (9) and (33) we obtain

$$|F_{i+1}u - F_{i+2}u| = |S^{-1}f_{i+1}|$$

$$= |(T_0 S)^{-1} T_0(f_{i+1} - f_i - S(F_{i+1}u - F_i u))|$$

$$\leq |(T_0 S)^{-1}| h_0(|F_{i+1}u - x| + |F_i u - x| + |y - x|) |F_{i+1}u - F_i u|$$

$$\leq \frac{w_i(t)}{2\phi(t)+q} (w_i(t) + 2(w_0(t) + \ldots + w_{i-1}(t)) + q)$$

$$= w_{i+1}(t).$$

(36)

In this manner we have established (35). Now we intend to show that $u \in Z(t)$ implies $(F_{m-1}u, F_m u) \in Z(\tilde{w}(t))$. It will suffice to prove the following inequalities:

$$|F_{m-1}u - F_m u| \leq w_{m-1}(t),$$ (37)

$$|F_m u - x_0| \leq s(t_0) - s_m(t),$$ (38)

$$|S_m^{-1} f_m| \leq w_m(t)$$ (39)

$$|(T_0 S_m)^{-1}| \leq h_0^{-1}(2\phi(\tilde{w}(t)) + w_{m-1}(t))^{-1}$$ (40)

The first inequality is a consequence of (35) and (38) follows from (36) for $i = m-1$. By (3.8'), (5) and (36) we have:

$$|T_0(S_m - S_0)| \leq |T_0(S_m - f'(x_0))| + |T_0(f'(x_0) - S_0)|$$

$$\leq h_0(|F_m u - x_0| + |F_{m-1}u - x_0| + |x_0 - x_{-1}|)$$

$$\leq h_0(2s(t_0) - 2s_m(t) + q_0 - w_{m-1}(t))$$

$$= 1 - h_0(2\phi(\tilde{w}(t)) + w_{m-1}(t)).$$

According to Lemma 4.10 this implies the invertibility of S_m and the inequality (40). Using the identity $f_{m-1} = -S(F_m u - F_{m-1} u)$ we obtain:

$$|S_m^{-1} f_m| = |(T_0 S_m)^{-1} T_0 (f_m - f_{m-1} - S(F_m u - F_{m-1} u))|$$

$$\leq h_0 |(T_0 S_m)^{-1}| (|F_m u - x| + |F_{m-1} u - x|$$

$$+ |y - x|) |F_m u - F_{m-1} u|$$

$$\leq \frac{w_{m-1}(t)}{2\phi(\tilde{w}(t)) + w_{m-1}(t)} (w_{m-1}(t) + 2(w_0(t) + \ldots + w_{m-2}(t)) + q) = w_m(t).$$

Until now we have proved that conditions (3.10), (3.11), (3.12) of Proposition 3.4 are satisfied. Our next task is to show that the inclusion

$$(x_{n-1}^{m-1}, x_{n-1}^m) \in Z(|x_{n-1}^{m-1} - x_{n-1}^m|, |x_{n-1}^m - x_n^1|) \tag{41}$$

holds for each $n = 1, 2, \ldots$.

According to (3.15) and (35) we already know that:

$$(x_{n-1}^{m-1}, x_{n-1}^m) \in Z(w_{m-1}^{(n-1)}(t_0), w_m^{(n-1)}(t_0)), \tag{42}$$

$$|x_{n-1}^{m-1} - x_{n-1}^m| \leq w_{m-1}^{(n-1)}(t_0), \tag{43}$$

$$|x_{n-1}^m - x_n^1| \leq w_m^{(n-1)}(t_0). \tag{44}$$

In view of the monotonicity of the function s it follows from the above relations that:

$$|x_{n-1}^m - x_0| \leq s(t_0) - s(|x_{n-1}^m - x_{n-1}^{m-1}|, |x_{n-1}^m - x_n^1|),$$

$$|(T_0 \delta f(x_{n-1}^{m-1}, x_{n-1}^m))^{-1}|$$

$$\leq h_0^{-1} (2\phi(|x_{n-1}^m - x_{n-1}^{m-1}|, |x_{n-1}^m - x_n^1|) + |x_{n-1}^m - x_{n-1}^{m-1}|)^{-1}.$$

The above inequalities together with (42) imply (41).

According to Proposition 3.4 there exists a point $x^* \in \bar{U}$ which is the common limit of the sequences $(x_n^j)_{n \geq 1}$ $(0 \leq j \leq m)$ and the estimates (31) and (32) are satisfied. Thus the proof of our theorem will be complete if we demonstrate that x^* is a root of the equation $f(x) = 0$. To show this we observe that it follows from (3) and (9) that

$$|T_0 f(x_{n+1}^1)| = |T_0(f(x_{n+1}^1) - f(x_n^m) - \delta f(x_n^{m-1}, x_n^m)(x_{n+1}^1 - x_n^m)|$$

$$\leq h_0(|x_{n+1}^1 - x_n^m| + |x_n^{m-1} - x_n^m|) \, |x_{n+1}^1 - x_n^m|$$

and then, using the continuity of f on $\bar{U}$, we conclude that $f(x^*) = 0$. □

We conclude this section by remarking that for the triplet (f, x_0, x_{-1}) considered in Proposition 4.4 the estimates (31) and (32) are attained for all $j = 0, 1, \ldots, m$ and $n = 1, 2, 3, \ldots$. In Section 6 we shall make some comments on the order of convergence and the efficiency of the iterative procedures (1), (2) and (3).

Comments

The method of nondiscrete mathematical induction was applied to the study of the modified secant method and of the secant method in [30] and [32]. In the latter paper multidimensional rates of convergence were introduced for the first time. The results obtained in these two papers were reproved in [34] for the case where consistent approximations of the derivative were used instead of divided differences.

Condition (7), which is shown to be optimal in Proposition 4.4, appears for the first time in [68] in the case of symmetric divided difference satisfying (13.3). In the case of consistent approximations of the derivative this condition appears for example in [6] and [60].

As far as we know the first semi-local analysis for the iterative procedure (3) was done by P. Laasonen [21] in case $m = 2$ and for divided differences satisfying condition (13.3). His result was improved in [39] where optimal convergence conditions as well as sharp error bounds were obtained. The generalization for an arbitrary m appeared in [42]. Finally, in [35] it is shown that the results of [42] remain valid if the divided differences are replaced by consistent approximations of the derivative.

A semi-local convergence theorem for an iterative procedure which generalizes the iterative procedure (3) was given by J.E. Dennis [9], (see also [37] and [38]).

5 Newton's method and some modifications

We have already studied Newton's method in Section 2 where we wanted to show, by a typical example, how nondiscrete induction can be applied to the investigation of iterative procedures. In what follows we shall present an improved version of the results presented there. Together with Newton's method we shall also study some iterative algorithms which can be obtained by modifying Newton's method in the same way as the iterative algorithms (4.2) and (4.3) have been obtained from the secant method. More precisely in the present section we are going to study the following iterative algorithms:

$$x_{n+1} = x_n - f'(x_n)^{-1} f(x_n), \quad n = 0,1,2,\ldots \tag{1}$$

$$x_{n+1} = x_n - f'(x_0)^{-1} f(x_n), \quad n = 0,1,2,\ldots \tag{2}$$

$$x_n^0 = x_{n-1}^m, \quad x_n^{j+1} = x_n^j - f'(x_n^0)^{-1} f(x_n^j); \tag{3}$$

$$j = 0,1,\ldots,m-1$$
$$n = 1,2,3,\ldots$$

The Newton method (1) is perhaps the best known iterative procedure for solving nonlinear equations. It is known that under some natural conditions it converges quadratically. The iterative algorithm (2) is usually called the simplified Newton method. It converges only linearly. In what follows the iterative algorithm (3) will be called Traub's method, because, as far as we know, J.F. Traub was the first to consider it in case of a scalar equation. He proved [67] that the order of convergence of this procedure equals $m + 1$.

The class $C'(k_0,r_0)$

The iterative algorithms (1), (2) and (3) presented above will be studied for pairs (f,x_0) belonging to the class $C'(k_0,r_0)$ defined as follows:

5.1 Definition Let $k_0 > 0$ and $r_0 \geq 0$ be two given real numbers. We say that a pair (f, x_0) belongs to the class $C'(k_0, r_0)$ if:

(c_1') f is a nonlinear operator defined on a subset D_f of a Banach space X and with values in a Banach space Y.

(c_2') x_0 is an interior point of D_f.

(c_3') f is Fréchet differentiable in the open ball $U = U(x_0, \mu)$ and continuous on its closure $\bar{U}$.

(c_4') The linear operator $S_0 = f'(x_0)$ is invertible, $T_0 = S_0^{-1} \in B(Y, X)$, and

$$|T_0(f'(x) - f'(y))| \leq k_0 |x - y| \text{ for all } x, y \in U. \tag{4}$$

(c_5') The following inequalities are satisfied:

$$|T_0 f(x_0)| \leq r_0, \tag{5}$$

$$2 k_0 r_0 \leq 1, \tag{6}$$

$$\mu \geq k_0^{-1}(1 - \sqrt{1 - 2 k_0 r_0}) =: \mu_0 \tag{7}$$

The conditions defining the class $C'(k_0, r_0)$ are in fact the hypotheses of the Kantorovich theorems in their affine invariant form. The affine invariance means here that if $(f, x_0) \in C'(k_0, r_0)$ and if $A \in B(Y, X)$ is invertible then $(Af, x_0) \in C'(k_0, r_0)$. The importance of having affine invariant conditions is stressed by Deuflhard and Heindl [11].

It is easy to see that $(f, x_0) \in C'(k_0, r_0)$ if and only if $(f, x_0, x_0) \in C(\frac{1}{2} k_0, 0, r_0)$, where $C(h_0, q_0, r_0)$ is the class defined in the preceding section. Using this observation, we obtain the following corollary of Theorem 4.6:

5.2 Theorem Let $k_0 > 0$ and $r_0 \geq 0$ be two real numbers satisfying condition (6) and denote

$$a = k_0^{-1}(1 - 2k_0 r_0)^{1/2}. \tag{8}$$

If (f, x_0) belongs to $C'(k_0, r_0)$ then the equation $f(x) = 0$ has a root x^*

belonging to the closed ball $W = \{x \in X; \; |x - x_0| \leq \mu_0\}$ *and this root is unique in the set* $V = \bar{U} \cap U(x_0, \mu_0 + 2a)$ *if* $a > 0$, *or in* W *if* $a = 0$.

The result stated above is sharp. Indeed, using Proposition 4.7, we obtain:

5.3 Proposition *Let* $k_0 > 0$ *and* $r_0 \geq 0$ *be two real numbers satisfying condition (6) and let* a *be the constant defined by (8).*

(i) *If* $a > 0$, *then there exist a function* $f: \mathbb{R} \to \mathbb{R}$, *and three points* x_0, x^*, $y^* \in \mathbb{R}$ *such that* $(f, x_0) \in C'(k_0, r_0)$, $f(x^*) = f(y^*) = 0$,

$$|x_0 - x^*| = \mu_0, \quad |x_0 - y^*| = \mu_0 + 2a.$$

(ii) *If* $a = 0$, *then for each* $\varepsilon > 0$ *there exists a function* $f_\varepsilon : \mathbb{R} \to \mathbb{R}$ *and three points* $x_0, x^*, y^*_\varepsilon$ *such that*

$$(f_\varepsilon, x_0) \in C'(k_0, r_0), \quad f_\varepsilon(x^*) = f_\varepsilon(y^*_\varepsilon) = 0,$$

$$|x^* - x_0| = \mu_0, \quad |y^*_\varepsilon - x_0| = \mu_0 + \varepsilon.$$

In the preceding section we made some comments on the inequality (4.7) stressing its special role among the conditions defining the class $C(h_0, q_0, r_0)$. Similar comments can be made on inequality (6) appearing in the definition of the class $C'(k_0, r_0)$. Taking $q_0 = 0$ in Proposition 4.8 we have

5.4 Proposition *Let* $k_0 > 0$ *and* $r_0 > 0$ *be two real numbers which do not satisfy condition (6). Then there exists a function* $f: \mathbb{R} \to \mathbb{R}$ *and a point* $x_0 \in \mathbb{R}$ *such that:*

(i) *conditions* $(c'_1) - (c'_4)$ *as well as inequality (5) are satisfied;*

(ii) *the equation* $f(x) = 0$ *has no solution.*

Finally let us note that from condition (4) it follows that for all $u, v \in U$.

$$|T_0(f(u) - f(v) - f'(v)(u - v))| \leq \frac{1}{2} k_0 \, |u - v|^2. \tag{9}$$

This inequality will be essentially used in what follows. A way of proving it is given in Section 2 or in Appendix B.

The simplified Newton method

The simplified Newton method (2) can be obtained as a particular case of the simplified secant method (4.2) by taking $x_0 = x_{-1}$. Thus, using again the observation that $(f,x_0) \in C'(k_0,r_0)$ if and only if $(f,x_0,x_0) \in C(\frac{1}{2}k_0, 0, r_0)$, it follows from the results obtained in Section 4 that:

5.5 __Theorem__ *If* $(f,x_0) \in C'(k_0,r_0)$ *then the iterative procedure (2) yields a sequence* $(x_n)_{n \geq 0}$ *of points belonging to the open sphere* $U(x_0,\mu_0)$ *which converges to a root* x^* *of the equation* $f(x) = 0$ *and the following estimates hold*

$$|x_n - x^*| \leq s(w^{(n)}(r_0)), \; n = 0,1,2,\ldots \tag{10}$$

$$|x_n - x^*| \leq s(|x_n - x_{n-1}|) - |x_n - x_{n-1}|, \; n = 1,2,3,\ldots \tag{11}$$

where w *and* s *are the functions defined by*

$$w(r) = \frac{1}{2} k_0 r^2 + r(1 - \sqrt{k_0^2 a^2 + 2k_0 r})$$

$$\tag{12}$$

$$s(r) = -a + \sqrt{a^2 + 2rk_0^{-1}}$$

with the constant à *given in (8).*

Newton's method

In Section 1 we have observed that for any given non-negative constant a the function

$$w(r) = \frac{1}{2} r^2 (r^2 + a^2)^{-1/2} \tag{13}$$

is a rate of convergence (of type (1.1)) on the interval $T = (0,\infty)$ and the corresponding estimate function is

$$s(r) = r - a + (r^2 + a^2)^{1/2}. \tag{14}$$

(See Example 1.4). In Section 2 we have shown that this rate of convergence appears in a natural way when studying Newton's method. It is interesting to note that for this rate of convergence we could obtain a nice explicit

81

expression for the quantities $s(w^{(n)}(r))$. A way of doing this is described in Appendix C. In what follows we content ourselves by giving these expressions in case the constant a is given by (8) and $r = r_0$.

5.6 Proposition *Let k_0 and r_0 be two positive real numbers satisfying condition (6), let a be the constant given by (8), and let w and s be the functions given by (13) and (14); set*

$$s_n = s(w^{(n)}(r_0)).$$

If $a > 0$ then for any non-negative integer n we have

$$s_n = \frac{2a\,\theta_0^{2^n}}{1 - \theta_0^{2^n}} \tag{15}$$

where $\theta_0 = \dfrac{1 - k_0(a + r_0)}{k_0 r_0}$ *, and if $a = 0$ then*

$$s_n = k_0^{-1} 2^{-n} \tag{16}$$

for $n = 0,1,2,\ldots$.

Proof It is easy to verify the identity

$$[s(r)]^2 = 2[s(r) + a]s(w(r)).$$

It follows that

$$s_{n+1} = \frac{s_n^2}{2(s_n + a)} \quad \text{for } n = 0,1,2,\ldots \tag{17}$$

If $a = 0$ then the above relation reduces to $s_{n+1} = \frac{1}{2} s_n$, whence we deduce the expression (16).

Suppose $a > 0$. In this case $0 < \theta_0 < 1$. For $n = 0$ (15) can be easily checked. If it holds for a certain $n \geq 0$, then using (17) we can write

$$s_{n+1} = \frac{4a^2\,\theta_0^{2^{n+1}}}{(1 - \theta_0^{2^n})^2} \cdot \frac{1 - \theta_0^{2^n}}{2a(1 + \theta_0^{2^n})} = \frac{2a\,\theta_0^{2^{n+1}}}{1 - \theta_0^{2^{n+1}}} .$$

The proof is complete. ◻

82

Now we can state our result concerning the convergence of Newton's method:

5.7 **Theorem** *If* $(f, x_0) \in C'(k_0, r_0)$ *then the iterative algorithm (1) yields a sequence* $(x_n)_{n \geq 0}$ *of points from the open ball* $U(x_0, \mu_0)$ *converging to a root* x^* *of the equation* $f(x) = 0$ *and the following estimates hold*

$$|x_n - x^*| \leq s_n \quad n = 0, 1, 2, \ldots \tag{18}$$

$$|x_n - x^*| \leq (|x_n - x_{n-1}|^2 + a^2)^{1/2} - a, \ n = 1, 2, \ldots \tag{19}$$

where a *is the constant given by (8) and* $(s_n)_{n \geq 0}$ *is the sequence defined in Proposition 5.6.*

Proof We shall apply Proposition 1.9. To this effect we denote by D the set of those points x of $U(x_0, \mu_0)$ for which the linear operator $f'(x)$ is invertible and we define the mapping

$$G : D \to X, \quad Gx = x - f'(x)^{-1} f(x)$$

Let w and s be the functions defined by (13) and (14). For any $r > 0$ we consider the set

$$Z(r) = \{x \in D; \ |x - x_0| \leq s(r_0) - s(r), \ |f'(x)^{-1} f(x)| \leq r\}. \tag{20}$$

We want to show that the hypotheses of Proposition 1.9 are satisfied. The relation $x_0 \in Z(r_0)$ is obvious so that all we have to prove is that $x \in Z(r)$ implies $Gx \in Z(w(r))$.

Set $z = Gx$. Using the identity $s(r) = r + s(w(r))$ we can write

$$|z - x_0| \leq |z - x| + |x - x_0| \leq r + s(r_0) - s(r)$$

$$= s(r_0) - s(w(r)).$$

We have $s(r_0) = \mu_0$ so that the above relation shows in particular that $z \in U(x_0, \mu_0)$. Then applying (4) we obtain

$$|T_0(S_0 - f'(z))| \leq k_0 |z - x_0| \leq k_0(\mu_0 - s(w(r)))$$

$$= 1 - k_0 \sqrt{r^2 + a^2}. \tag{21}$$

83

According to Lemma 4.10 it follows that the linear operator $f'(z)$ is invertible and

$$|(T_0 f'(z))^{-1}| \leq k_0^{-1}(r^2 + a^2)^{-1/2}. \tag{22}$$

The relation defining Gx gives

$$f(x) + f'(x)(z - x) = 0,$$

so that using (9) we obtain:

$$|T_0 f(z)| = |T_0(f(z) - f(x) - f'(x)(z - x))|$$

$$\leq \frac{1}{2} k_0 |z - x|^2 \leq \frac{1}{2} k_0 r^2. \tag{23}$$

Finally, according to (22) and (23), we have the inequality

$$|f'(z)^{-1} f(z)| = |(T_0 f'(z))^{-1} T_0 f(z)|$$

$$\leq \frac{1}{2} r^2(r^2 + a^2)^{-1/2} = w(r),$$

which together with (21) shows that $z \in Z(w(r))$. Thus the hypotheses of Proposition 1.9 are verified. Hence the sequence $(x_n)_{n \geq 0}$ converges to a point x^* and the estimates (18) are verified. Moreover for all n we shall have $x_{n-1} \in Z(w^{(n-1)}(r_0))$ and $|x_n - x_{n-1}| \leq w^{(n-1)}(r_0)$. The function s being increasing it will follow that $x_{n-1} \in Z(|x_n - x_{n-1}|)$ and thus by virtue of point 3. of Proposition 1.9 the estimates (19) will also be satisfied.

Let us observe now that by taking $x = x_{n-1}$ and $z = x_n$ in (23) we get $|T_0 f(x_n)| \leq \frac{1}{2} k_0 |x_n - x_{n-1}|^2$; using the continuity of f on $\bar{U}$ we deduce that $f(x^*) = 0$.

This completes the proof of our theorem.   □

Traub's method

Traub's method (3) can obviously be interpreted as an iterative algorithm of type (1.m) in the sense specified in Section 3. We shall associate with this algorithm a rate of convergence of type (1.m). In order to be able to treat the cases $m = 1$ and $m \geq 1$ together we adopt the convention made before the statement of Lemma 4.12.

5.8 **Lemma** *Let* T *denote the real positive semiaxis. Suppose we are given a non-negative real constant* a *and a positive integer* m. *For all* $r \in T$ *consider the functions*

$$\phi(r) = r + \sqrt{r^2 + a^2}, \quad w_0(r) = r$$

and define recursively

$$w_{k+1}(r) = \frac{w_k(r)}{2\phi(r)} \; (2(w_0(r) + \ldots + w_{k-1}(r)) + w_k(r)), \quad k = 0, 1, \ldots, m-2$$

$$w_m(r) = w_{m-1}(r) \cdot \frac{2(w_0(r) + \ldots + w_{m-2}(r)) + w_{m-1}(r)}{2(\phi(r) - w_0(r) - \ldots - w_{m-1}(r))}.$$

Then the mapping $w(r) = (w_1(r), \ldots, w_m(r))$ *is a rate of convergence of type* $(1.m)$ *on* T *and the corresponding estimate function is given by*

$$s(r) = \phi(r) - a.$$

<u>Proof</u> Apply Proposition 3.3.　　　　　　　　　　　　　　　　　　□

Using the above lemma and Proposition 3.4 we can prove the following theorem.

5.9 **Theorem** *Consider the rate of convergence obtained in Lemma 5.8, with the constant* a *given by (8), and let* s_j, $0 \leq j \leq m$, *be the mappings attached to it by (3.9). If* $(f, x_0) \in C'(k_0, r_0)$, *then the iterative algorithm (3), with starting point* $x_0^m = x_0$, *produces* $m + 1$ *sequences* $(x_n^j)_{n \geq 0}$, $0 \leq j \leq m$, *which converge to a root* x^* *of the equation* $f(x) = 0$ *and the following estimates hold for all* $j = 0, 1, \ldots, m$ *and* $n = 1, 2, 3, \ldots$

$$|x_n^j - x^*| \leq s_j(w_n^{(n-1)}(r_0)), \tag{24}$$

$$|x_n^j - x^*| \leq s_j(|x_n^1 - x_n^0|). \tag{25}$$

<u>Proof</u> Let $u \in U$. Set $F_0 u = u$ and define recursively

85

$$F_{j+1}u = F_ju - f'(u)^{-1}f(F_ju), \quad j = 0,1,\ldots,m-1 \tag{26}$$

Let us denote by D the set of those $u \in U$ for which the above formulae make sense, i.e. $f'(u)$ is invertible and $F_ku \in U$ for all $k = 0,1,\ldots,m-1$. Let $F:D \to X^{m+1}$ be the mapping which transforms u into $(F_0u, F_1u,\ldots,F_mu)$. This mapping is obviously of the form (3.4) with $p = 1$. In the case $p = 1$ we have $P = P_m = $ the projection onto the last component. If we consider the sequences generated by (3.5) and (3) then we have:

$$x_n^j = P_j z_n; \quad j = 0,1,\ldots,m, \quad n = 1,2,3,\ldots$$

For each $t > 0$ let us define the set

$$Z(t) = \{u \in X; \; |u - x_0| \le s(r_0) - s(t), \tag{27}$$

the linear operator $S = f'(u)$ is invertible and $|S^{-1}f(u)| \le t\}$.

We have obviously $Z(t) \subset U$ because $s(r_0) = \mu_0 \le \mu$. Let us prove that $u \in Z(t)$ implies that the following relations hold for $k = 0,1,\ldots,m-1$:

$$F_ku \in U, \tag{28}$$

$$|F_ku - F_{k+1}u| \le w_k(t). \tag{29}$$

For $k = 0$ relation (28) becomes $u \in U$ while relation (29) reduces to $|S^{-1}f(u)| \le t$, both of them being thus immediate consequences of the hypothesis $u \in Z(t)$. Now let us suppose that (28) and (29) hold for $k=0,1,\ldots,i$ where $0 \le i \le m-2$. Then we have

$$|F_{i+1}u - x_0| \le |u - x_0| + \sum_{j=0}^{i} |F_{j+1}u - F_ju|$$

$$\le s(r_0) - s(t) + \sum_{j=0}^{i} w_j(t) = s(r_0) - s_{i+1}(t). \tag{30}$$

In particular it follows that (28) holds for $k = i + 1$ too. We shall prove that the same is true for (29). First of all let us note that (4) implies

$$|T_0(S_0 - S)| \le k_0 |u - x_0| \le k_0(s(r_0) - s(t)) = 1 - k_0 \phi(t).$$

Then, applying Lemma 4.10, we deduce that the linear operator $T_0S = I-T_0(S_0-S)$ is invertible and

$$|(T_0 S)^{-1}| \le \frac{1}{k_0 \, \phi(t)} \; . \tag{31}$$

If we denote $f_j = f(F_j u)$, then from (26) we have

$$f_{i+1} = f_{i+1} - f_i - S(F_{i+1}u - F_i u).$$

Denoting now $S_i = f'(F_i u)$ and using (4) and (9) we may write

$$|T_0 f_{i+1}| \le |T_0 (f_{i+1} - f_i - S_i (F_{i+1}u - F_i u))|$$

$$+ \; |T_0 (S_i - S)(F_{i+1}u - F_i u)|$$

$$\le \frac{1}{2} k_0 |F_{i+1}u - F_i u|^2 + k_0 |F_i u - u| \; |F_{i+1}u - F_i u|$$

$$\le \frac{1}{2} k_0 w_i(t)(2(w_0(t) + \ldots + w_{i-1}(t)) + w_i(t)). \tag{32}$$

Finally from (31) and (32) we obtain

$$|F_{i+1}u - F_{i+2}u| = |S^{-1}f_{i+1} = |(T_0 S)^{-1}T_0 f_{i+1}| \le w_{i+1}(t).$$

Thus it follows that the relations (28) and (29) are satisfied for all $k = 0,1,\ldots,m-1$. From (28) it results that $Z(t) \subset D$.

On the other hand we have $Z(r_0) = \{x_0\}$ so that condition (3.10) of Proposition 3.4 is satisfied. Condition (3.12) is also satisfied because it coincides with (29). Let us prove now that condition (3.11) holds too. It is sufficient to prove that $u \in Z(t)$ implies the following two statements

$$|F_m u - x_0| \le s(r_0) - s(w_m(t)), \tag{33}$$

$$S_m = f'(F_m u) \text{ is invertible and } |S_m^{-1}f(F_m u)| \le w_m(t). \tag{34}$$

Inequality (33) follows immediately by observing that (30) holds for $i = m-1$ and that $s(w_m(t)) = s_m(t)$. In order to prove (34) let us note that (4) and (33) imply

$$|T_0(S_m - S_0)| \le k_0 |F_m u - x_0|$$

$$\le 1 - k_0(\phi(t) - w_0(t) - \ldots - w_{m-1}(t)).$$

Inequality (32) holds for i = m-1 too so that we have

$$|S_m^{-1} f_m| = |(I + T_0(S_m - S_0))^{-1} T_0 f_m|$$

$$\leq w_{m-1}(t) \frac{w_{m-1}(t) + 2(w_0(t) + \ldots + w_{m-2}(t))}{2(\phi(t) - w_0(t) - \ldots - w_{m-1}(t))} = w_m(t).$$

The hypotheses of Proposition 3.4 being satisfied it follows that there exists a point $x^* \in X$ which is the common limit of the sequences $(x_n^j)_{n \geq 1}$, $0 \leq j \leq m$ and that the estimates (24) as well as the relation $x_{n-1}^m \in Z(w_m^{(n-1)}(r_0))$ hold for all positive integers n. Using the fact that the function s is increasing on $(0, \infty)$ from this relation we can deduce that $x_{n-1}^m \in Z(|x_n^1 - x_{n-1}^m|)$. According to point 4. of Proposition 3.4 this implies the estimates (25).

Taking i = 0 and u = x_{n-1}^m in (32) we obtain

$$|T_0 f(x_n^1)| \leq \frac{1}{2} k_0 |x_n^1 - x_n^0|^2 + k_0 |x_n^1 - x_n^0|^2.$$

Using the continuity of f on $\bar{U}$ we deduce that $f(x^*) = 0$. The proof is complete. □

Sharpness of the estimates

In Theorems 5.5, 5.7, 5.9 we have shown that the simplified Newton method, the Newton method and Traub's method may be considered as iterative procedures for the class $C'(k_0, r_0)$ (the first two of type (1.1), the last type (1.m)) in the sense of Definition 3.6. Using the terminology introduced in Section 3 we can say that inequalities (10), (18) and (24) are apriori error estimates, while inequalities (11), (19) and (25) are aposteriori error estimates for these iteraative procedures. The aposteriori error estimates are generally better than the apriori ones. However it turns out that all these estimates are sharp. Indeed we have

5.10 **Proposition** *If $k_0 > 0$ and $r_0 \geq 0$ are two real numbers satisfying inequality (6), then there exist a function $f: \mathbb{R} \to \mathbb{R}$ and a point $x_0 \in \mathbb{R}$ such that the pair (f, x_0) belongs to $C'(k_0, r_0)$ and for this pair the estimates (10), (11), (18), (19), (24) and (25) are attained for all n and j.*

<u>Proof</u> Take $f(x) = x^2 - a^2$, where a is given by (8), $x_0 = k_0^{-1}$, $\mu = 2r_0$ and apply Proposition 3.7. (See also the proof of Proposition 4.4). □

6 Rates and orders of convergence

Introduction

The notion of order of convergence has been introduced because of the necessity of comparing different iterative procedures. The first definition of this notion was given in 1870 by E. Schröder [62]. He considered iterative procedures of the form

$$x_{n+1} = F(x_n), \quad n = 0,1,2,\ldots \tag{1}$$

where F was a mapping of the complex plane into itself. Supposing that the sequence $(x_n)_{n \geq 1}$ defined by (1) converges towards a point x^* and that F was analytic in a neighbourhood of x^*, he said that the iterative procedure had the order of convergence k if

$$F^{(j)}(x^*) = 0, \quad j = i,2,\ldots,k-1; \quad F^{(k)}(x^*) \neq 0. \tag{2}$$

It is easy to see that under these assumptions there exist two positive constants A and B and a natural number N such that

$$A|x_n - x^*|^k \leq |x_{n+1} - x^*| \leq B|x_n - x^*|^k \quad \text{for } n \geq N. \tag{3}$$

Let us clear up the meaning of the above relation from a numerical point of view. In what follows we suppose that for all $n \geq N$ we have $0 < |x_n - x^*| < 1$. This assumption is quite natural because on the one hand we have clearly $|x_n - x^*| < 1$ for sufficiently large n and, on the other hand, if for a certain m $|x_n - x^*| = 0$ then, according to (3), $|x_n - x^*| = 0$ for all $n \geq m$. If $(x_n)_{n \geq 0}$ is a sequence of real numbers and if $0 < |x_n - x'| < 1$, then $[\log_{0,1}|x_n - x^*|] + 1$ represents the number of "exact decimals" of x_n (if x is a positive real number then [x] denotes the greatest integer less than or equal to x.) From (3) it follows that

$$\log_{0,1}B + k \log_{0,1} |x_n - x^*| \leq \log_{0,1}|x_{n+1} - x^*|$$

$$\leq \log_{0,1} A + k \log_{0,1}|x_n - x^*|.$$

If n is large then the constants $\log_{0.1} A$ and $\log_{0.1} B$ are small in comparison with $\log_{0.1} |x_n - x^*|$ and $\log_{0.1} |x_{n+1} - x^*|$ so that for sufficiently large n we have

$$\log_{0.1} |x_{n+1} - x^*| \approx k \log_{0.1} |x_n - x^*|.$$

This means that, for sufficiently large n, at each step of the iterative procedure the number of "exact decimals" multiplies by exactly k. If instead of (3) we had only

$$|x_{n+1} - x^*| \le B|x_n - x^*|^k \text{ for } n \ge N, \tag{4}$$

then this would imply that the number of exact decimals multiplies at least k times at each step of the iterative procedure.

In the same paper E. Schröder considered also iterative procedures of the form (1) where F satisfied the condition

$$0 < |P'(x^*)| < 1. \tag{5}$$

He said that such iterative procedures were of order one. It is clear that a sequence produced by an iterative procedure of order one converges more slowly than an iterative procedure of order k ($k \ge 2$). Indeed from (5) it follows that there exist two real numbers a and b with $0 < a \le b < 1$ and a natural number N such that

$$a|x_n - x^*| \le |x_{n+1} - x^*| \le b|x_n - x^*|, \, n \ge N. \tag{6}$$

This shows that the sequence $(|x_n - x^*|)_{n \ge N}$ behaves like a geometric progression. By analogy with (4) we can consider also the weaker condition

$$|x_{n+1} - x^*| \le b|x_n - x^*|, \, n \ge N. \tag{7}$$

which says that the sequence $(x_n)_{n \ge N}$ converges at least as fast as a geometric progression.

We note that the conditions (3), (4), (6) and (7) as opposed to conditions (2) and (5), do not depend on the particular form of the iterative procedure. Moreover the number k appearing in (3) and (4) does not need to be an integer, but it can be any real number (for obvious reasons we are interested only in the case where k is greater than one). These remarks lead to the definition of the notion of Q-order of convergence of a sequence to be given in the next paragraph:

Q-orders of convergence

6.1 Definition *Let $(x_n)_{n \geq 0}$ be a sequence of points of a metric space (X,d) which converges to a point $x^* \in X$.*

(i) *We say that 1 is a Q-order of convergence of the sequence $(x_n)_{n \geq 0}$ if there exists a constant $b \in (0,1)$ and a positive integer N such that*

$$d(x_{n+1}, x^*) \leq b \, d(x_n, x^*) \text{ for } n \geq N. \tag{8}$$

(ii) *We say that 1 is the exact Q-order of convergence of the sequence $(x_n)_{n \geq 0}$ if there exist two constants a, $b \in (0,1)$ and a positive integer N such that*

$$a \, d(x_n, x^*) \leq d(x_{n+1}, x^*) \leq b \, d(x_n, x^*) \text{ for } n \geq N. \tag{9}$$

(iii) *We say that the number $r \in (1, +\infty)$ is a Q-order of convergence of the sequence $(x_n)_{n \geq 0}$ if there exist a constant $B > 0$ and a natural number N such that*

$$d(x_{n+1}, x^*) \leq B \, d(x_n, x^*)^r \text{ for } n \geq N. \tag{10}$$

(iv) *We say that the number $r \in (1, +\infty)$ is the exact Q-order of convergence of the sequence $(x_n)_{n \geq 0}$ if there exist two positive constants A and B and a natural number N such that*

$$A \, d(x_n, x^*)^r \leq d(x_{n+1}, x^*) \leq B \, d(x_n, x^*)^r \text{ for } n \geq N. \tag{11}$$

It is easy to see that if a number $r \in [1, +\infty)$ is a Q-order of convergence of the sequence $(x_n)_{n \geq 0}$ then either

$$d(x_n, x^*) > 0 \text{ for all } n \geq N, \tag{12}$$

or there exist a positive integer m such that

$$d(x_n, x^*) = 0 \text{ for all } n \geq m. \tag{13}$$

We are not interested in the latter situation so that from now on we shall suppose that (12) is satisfied. We note that if a sequence $(x_n)_{n\geq 0}$ has a Q-order of convergence then the condition

$$d(x_n, x_{n+1}) > 0, \quad n \geq N \tag{14}$$

implies condition (12).

If condition (12) is satisfied and if r is the exact Q-order of convergence of the sequence $(x_n)_{n\geq 0}$ then r is the unique number with this property. On the other hand if $r \in [1,\infty)$ is a Q-order of convergence of a sequence $(x_n)_{n\geq 0}$ then any number r_1, with $1 \leq r_1 \leq r$ is also a Q-order of convergence of this sequence. Denoting by $Q(x_n)$ the set of all Q-orders of convergence of the sequence $(x_n)_{n\geq 0}$ it follows that $Q(x_n)$ is nonvoid if and only if $1 \in Q(x_n)$. Suppose $Q(x_n)$ is nonvoid and set

$$\rho = \sup Q(x_n).$$

It follows that

$$[1,\rho) \subset Q(x_n) \subset [1,\rho].$$

It is clear that if the exact Q-order of convergence of a sequence $(x_n)_{n\geq 1}$ exists then it equals ρ and in this case we have $Q(x_n) = [1,\rho]$. Let us consider now the sequence of real numbers defined by

$$x_1 = a, \quad 0 < a < 1 < r, \quad x_{n+1} = x_n^{r - \frac{1}{n}} \quad n = 1,2,\dots. \tag{15}$$

For this sequence we have obviously $Q(x_n) = [1,r)$. On the other hand for the sequence $(x_n)_{n\geq 0}$ defined by

$$0 < a < 1 < r < s, \quad x_{2k} = a^{(rs)^k}, \quad x_{2k+1} = a^{r^k s^{k+1}}, \tag{16}$$

we have $Q(x_n) = [1,r]$, but this sequence does not possess an exact Q-order of convergence.

The above observations motivate the following definition:

6.2 **Definition** *Let* $(x_n)_{n\geq 0}$ *be a convergent sequence in a metric space* (X,d) *and let* $Q(x_n)$ *be the set of all Q-orders of convergence of this sequence. Then the number* $\rho = \sup Q(x_n)$ *is called the weak Q-order of convergence of the sequence* $(x_n)_{n\geq 0}.$

While constructing a sequence $(x_n)_{n \geq 0}$ by an iterative procedure we do not know in general, the limit x^* of this sequence. Thus we are not able to measure the quantities $d(x_n, x^*)$ appearing in Definition 6.1 at any finite stage of the process. However, it turns out that, if the sequence $(x_n)_{n \geq 0}$ converges sufficiently fast then, asymptotically, we can replace the quantity $d(x_n, x^*)$ by $d(x_{n+1}, x_n)$. Before giving a precise statement of this result let us fix the terminology:

If 1 is the exact Q-order of convergence of a sequence then we say that this sequence converges (exactly) linearly.

If 1 is a Q-order of convergence of a sequence then we say that this sequence converges at least linearly.

Let us suppose that condition (12) is satisfied. Then the sequence $(x_n)_{n \geq 0}$ converges at least linearly if and only if

$$\lim \sup \frac{d(x_{n+1}, x^*)}{d(x_n, x^*)} < 1.$$

The sequence $(x_n)_{n \geq 0}$ converges at least linearly but not exactly linearly if and only if

$$\lim \sup \frac{d(x_{n+1}, x^*)}{d(x_n, x^*)} < 1 \text{ and } \lim \inf \frac{d(x_{n+1}, x^*)}{d(x_n, x^*)} = 0.$$

The above condition is clearly satisfied if

$$\lim \frac{d(x_{n+1}, x^*)}{d(x_n, x^*)} = 0. \tag{17}$$

6.3 **Definition** *Let $(x_n)_{n \geq 0}$ be a sequence of points of a metric space (X, d) which converges to a point $x^* \in X$ such that condition (12) is fulfilled. We say that this sequence converges superlinearly if relation (17) is satisfied.*

It is easy to see that, if the sequence $(x_n)_{n \geq 0}$ has a Q-order of convergence greater than 1, then it converges superlinearly. The converse is not true. Indeed, let us consider the sequence of positive real numbers defined by

$$x_0 = a, \ 0 < a < 1/e, \ x_{n+1} = -\frac{x_n}{\log_e x_n}, \quad n = 0, 1, 2, \dots . \tag{18}$$

This sequence converges superlinearly to zero. Suppose now that there exists a $r \in (1, +\infty) \cap Q(x_n)$. This means that

94

$x_{n+1} \leq A x_n^r$ for sufficiently large n.

Hence

$$\frac{(1/x_n)^{r-1}}{\log_e (1/x_n)} \leq A \qquad (19)$$

for sufficiently large n which contradicts the well known fact that

$$\lim_{t \to \infty} t^{\varepsilon}/\log_e t = \infty \text{ for any } \varepsilon > 0.$$

It follows that $Q(x_n) = \{1\}$. Thus the class of superlinearly convergent sequences is effectively larger than the class of sequences having a Q-order of convergence greater than 1. For superlinearly convergent sequences we have the following two important results:

6.4 <u>Proposition</u> *Let* $(x_n)_{n \geq 0}$ *be a sequence of points of a metric space* (X,d) *which converges to a point* $x^* \in X$ *and suppose that*

$$d(x_n, x_{n+1}) > 0 \text{ for sufficiently large n.}$$

Then the sequence $(x_n)_{n \geq 0}$ *converges superlinearly if and only if the sequence* $(d(x_n, x_{n+1}))_{n \, 0}$ *converges superlinearly.*

<u>Proof</u> Suppose $(x_n)_{n \geq 0}$ converges superlinearly. Then there exists a positive integer N such that

$$d(x_{n+1}, x^*) \leq \frac{1}{2} d(x_n, x^*) \text{ for } n \geq N. \qquad (20)$$

It follows that for $n \geq N$, we have

$$\frac{d(x_{n+1}, x_{n+2})}{d(x_n, x_{n+1})} \leq \frac{d(x_{n+1}, x^*) + d(x_{n+2}, x^*)}{d(x_n, x^*) - d(x_{n+1}, x^*)} \leq 3 \frac{d(x_{n+1}, x^*)}{d(x_n, x^*)}$$

from which we deduce the superlinear convergence of the sequence $(d(x_n, x_{n+1}))_{n \geq 0}$.

Conversely, if we suppose that the sequence $(d(x_n, x_{n+1}))_{n \geq 0}$ converges superlinearly then there exists a positive integer N such that

$$d(x_{j+1}, x_j) \leq \frac{1}{3} d(x_j, x_{j-1}) \quad j \geq N.$$

For any $n \geq N$ we have

$$d(x_{n+1}, x^*) \leq \sum_{j=1}^{\infty} d(x_{n+j}, x_{n+j+1})$$

$$\leq d(x_{n+1}, x_{n+2}) \sum_{j=0}^{\infty} 3^{-j} = \frac{3}{2} d(x_{n+1}, x_{n+2})$$

$$\leq \frac{1}{2} d(x_n, x_{n+1}).$$

Finally, writing

$$\frac{d(x_{n+1}, x^*)}{d(x_n, x^*)} \leq \frac{\frac{3}{2} d(x_{n+1}, x_{n+2})}{d(x_n, x_{n+1}) - d(x_{n+1}, x^*)} \leq 3 \frac{d(x_{n+1}, x_{n+2})}{d(x_n, x_{n+1})}$$

we infer that the sequence $(x_n)_{n\geq 0}$ converges superlinearly. □

6.5 Proposition (Dennis-More [10]). *Let* $(x_n)_{n\geq 0}$ *be a sequence of points of a metric space* (X,d) *such that* $d(x_n, x_{n+1}) > 0$ *for sufficiently large* n. *If this sequence converges superlinearly to a point* $x^* \in X$ *then*

$$\lim_{n \to \infty} \frac{d(x_n, x_{n+1})}{d(x_n, x^*)} = 1.$$

Proof The result follows immediately from the inequality

$$\left| \frac{d(x_n, x_{n+1})}{d(x_n, x^*)} - \frac{d(x_n, x^*)}{d(x_n, x^*)} \right| \leq \frac{d(x_{n+1}, x^*)}{d(x_n, x^*)} \ .$$

□

Using the last two propositions we obtain the following result:

6.6 Corollary *Let* $(x_n)_{n\geq 0}$ *be a convergent sequence satisfying condition* (14) *and let* r *be a real number greater than one. Then:*

(i) r *is a Q-order of convergence of the sequence* $(x_n)_{n\geq 0}$ *if and only if it is a Q-order of convergence of the sequence* $(d(x_n, x_{n+1}))_{n\geq 0}$.

(ii) r *is the exact Q-order of convergence of the sequence* $(x_n)_{n\geq 0}$ *if and only if it is the exact Q-order of convergence of the sequence* $(d(x_n, x_{n+1}))_{n\geq 0}$.

<u>Proof</u> Writing

$$\frac{d(x_{n+1},x_{n+2})}{d(x_{n+1},x_n)^r} = \frac{d(x_{n+1},x_{n+2})}{d(x_{n+1},x^*)} \cdot \frac{d(x_n,x^*)^r}{d(x_{n+1},x_n)^r} \cdot \frac{d(x_{n+1},x^*)}{d(x_n,x^*)^r}$$

and

$$\frac{d(x_{n+1},x^*)}{d(x_n,x^*)^r} = \frac{d(x_{n+1},x^*)}{d(x_{n+1},x_{n+2})} \cdot \frac{d(x_n,x_{n+1})^r}{d(x_n,x^*)^r} \cdot \frac{d(x_{n+1},x_{n+2})}{d(x_n,x_{n+1})^r}$$

it follows that

$$\limsup \frac{d(x_{n+1},x^*)}{d(x_n,x^*)^r} = \limsup \frac{d(x_{n+1},x_{n+2})}{d(x_n,x_{n+1})^r}$$

and

$$\liminf \frac{d(x_{n+1},x^*)}{d(x_n,x^*)^r} = \liminf \frac{d(x_{n+1},x_{n+2})}{d(x_n,x_{n+1})^r} \ . \qquad \square$$

Let us remark that there are examples which show that, in the case $r = 1$, neither (i) nor (ii) holds. However, if one of the sequences

$$(x_n)_{n\geq 0}, \ (d(x_n,x_{n+1}))_{n\geq 0}$$

converges superlinearly then so does the other and

$$Q(x_n) = Q(d(x_n,x_{n+1})).$$

Rates of convergence

In what follows we shall assume that the metric space (X,d) is complete. Under this assumption it is easy to see that if $(x_n)_{n\geq 0}$ is a sequence of points of X such that the sequence $(d(x_n,x_{n+1}))_{n\geq 0}$ converges to zero at least linearly then the sequence $(x_n)_{n\geq 0}$ is convergent. According to Corollary 6.6 a number $r \in (1,\infty)$ is the Q-order of convergence of a sequence $(x_n)_{n\geq 0}$ if and only if there exists a positive constant B and a natural number N such that

$$d(x_n,x_{n+1}) \leq Bd(x_{n-1},x_n)^r, \quad n \geq N. \tag{21}$$

It also follows that r is the exact Q-order of convergence of the sequence $(x_n)_{n\geq 0}$ if and only if there exist two positive constants A and B and a natural number N such that

$$Ad(x_{n-1},x_n)^r \leq d(x_n,x_{n+1}) \leq Bd(x_{n-1},x_n)^r, \quad n \geq N. \tag{22}$$

The above inequalities, by contrast to (10) and (11), relate quantities which can be measured at finite stages of the process. The constants A, B and N can usually be computed from theoretical considerations. In many cases the estimates given by (21) and (22) are accurate only for very large N. We might want, however, to stop the process before the bound N is reached. Of course, it is possible to extend the validity of (21) and (22) for all n, by taking B sufficiently large and A sufficiently small, but this would give pessimistic estimates. The following example shows that we cannot expect, in general, to find estimates of the form (21) or (22) that are accurate for the whole length of the process.

Consider the quadratic polynomial $f(x) = x^2 - a^2$ where a is a positive constant and apply Newton's method for the equation $f(x) = 0$. For any starting point $x_0 > a$ we shall obtain a sequence $(x_n)_{n\geq 0}$ which converges monotonically to a. This sequence is given by the following recurrence scheme

$$x_{n+1} = \frac{x_n^2 + a^2}{2x_n}, \quad n = 0,1,2,\ldots .$$

It is easy to check that

$$|x_{n+1} - x_n| = w(|x_n - x_{n-1}|) \quad n = 0,1,2,\ldots$$

where

$$w(t) = \frac{t^2}{2\sqrt{t^2 + a^2}}. \tag{23}$$

In Section 1 we have seen that the function w defined above is a small function in the sense of Definition 1.1 and that the corresponding estimate function is $s(t) = t - a + \sqrt{t^2 + a^2}$. A closer inspection of formula (23) shows that, for very small t, the function w assumes approximately the form $\frac{1}{2a} t^2$ whereas, for large t, the summand t^2 predominates in the denominator, so that the function is approximately linear, $\frac{1}{2} t$.

Thus in the initial stages of the process, when $|x_n - x_{n-1}|$ is still large (we note that for any fixed m, $\lim\limits_{x_0 \to \infty} |x_m - x_{m-1}| = +\infty$), the relation between $|x_n - x_{n-1}|$ and $|x_{n+1} - x_n|$ is approximately linear, while asymptotically, in other words for small $|x_n - x_{n-1}|$, this relation becomes quadratic. It follows that accurate estimates valid for the whole process - including the initial steps - cannot be based on any simple quadratic monomial. However, using the function w given by (23) we have an exact description of the relation between $|x_n - x_{n+1}|$ and $|x_{n-1} - x_n|$ for the whole length of the process. This motivates the following definition:

6.7 <u>Definition</u> *Let* $(x_n)_{n \geq 0}$ *be a sequence of points of a complete metric space* (X,d) *satisfying condition (14) and let* w:T → T *be a small function in the sense of Definition 1.1, where* T *is an interval of the real line such that*

$$(0, \sup \{d(x_n, x_{n+1}); n \in \mathbb{Z}_+\}] \subset T \subset (0, +\infty) \tag{24}$$

Then

(i) *We say that* w *is a rate of convergence of the sequence* $(x_n)_{n \geq 0}$ *if*

$$d(x_{n+1}, x_n) \leq w(d(x_n, x_{n-1})) \text{ for } n = 1,2,3,\ldots \tag{25}$$

(ii) *We say that* w *is an exact rate of convergence of the sequence* $(x_n)_{n \geq 0}$ *if*

$$d(x_{n+1}, x_n) = w(d(x_n, x_{n-1})) \text{ for } n = 1,2,3,\ldots . \tag{26}$$

Let us note that if a number $r \in (1, +\infty)$ is a Q-order of convergence of a sequence, then the function $w(t) = ct^r$ is a rate of convergence of this sequence, for a suitable constant c. Conversely if a function w is a rate of convergence of a sequence $(x_n)_{n \geq 0}$ and if

$$\limsup_{t \downarrow 0} \frac{w(t)}{t^r} < \infty \tag{27}$$

then r is a Q-order of convergence of this sequence. Thus the notion of rate of convergence is a refinement of the notion of Q-order of convergence. In the definition of the notion of a rate of convergence it is not required that the rate of convergence should be a nondecreasing function. However the usual rates of convergence do have this property. For such rates of convergence we have the following result:

6.8 <u>Proposition</u> *Let* $(x_n)_{n \geq 0}$ *be a sequence of points of a complete metric space* (X,d). *If this sequence has a nondecreasing rate of convergence* w *then it converges to a point* $x^* \in X$, *and*

$$d(x_n,x^*) \leq s_1(d(x_n,x_{n-1})) \leq s(w^{(n)}(d(x_1,x_0))), \quad n = 1,2,3,\ldots \qquad (28)$$

where s *is the estimate function corresponding to the rate of convergence* w *and* $s_1(t) = s(t) - t$.

<u>Proof</u> By induction it follows that

$$d(x_{n+1},x_n) \leq w^{(n)}(d(x_1,x_0)), \quad n = 0,1,2,\ldots \ .$$

Thus the sequence $(x_n)_{n \geq 0}$ is convergent because the distances between its consecutive terms are majorized by the terms of a convergent series and the space X is complete. Finally, denoting by x^* the limit of the sequence $(x_n)_{n \geq 0}$, we have:

$$d(x_n,x^*) \leq \sum_{j=0}^{\infty} d(x_{n+j},x_{n+j+1}) \leq \sum_{j=1}^{\infty} w^{(j)}(d(x_n,x_{n-1}))$$

$$= s_1(d(x_n,x_{n-1})) \leq s_1(w^{(n-1)}(d(x_1,x_0))) = s(w^{(n)}(d(x_1,x_0))). \qquad \square$$

If F is a mapping of a complete metric space into itself such that

$$d(F(Fx), Fx) \leq w(d(Fx,x)), \quad x \in X,$$

where w is a small function, then, for any starting point x_0, the sequence $(x_n)_{n \geq 0}$ produced by the iterative algorithm

$$x_{n+1} = F(x_n) \quad n = 0,1,2,\ldots$$

has obviously the rate of convergence w (see also Proposition 1.13).
 However for sequences generated by iterative procedures of the form

$$x_{n+1} = F(x_{n-p+1}, x_{n-p},\ldots,x_n), \quad n = 0,1,\ldots \qquad (29)$$

it is not very natural to look for estimates of type (25) or (26). For example let us consider again the quadratic polynomial $f(x) = x^2 - a^2$, $a > 0$, and let $(x_n)_{n \geq 0}$ be the sequence obtained by applying the secant method to the equation $f(x) = 0$, i.e.

$$x_{n+1} = \frac{a^2 + x_n x_{n-1}}{x_n + x_{n-1}}, \quad n = 0,1,2,\ldots .$$

For any starting points $a < x_0 < x_{-1}$ the sequence $(x_n)_{n \geq -1}$ converges monotonically to a and we have

$$|x_{n+1} - x_n| = w(|x_n - x_{n-1}|, |x_{n-1} - x_{n-2}|), \quad n = 1,2,3,\ldots$$

where

$$w(t_1,t_2) = \frac{t_2(t_1 + t_2)}{t_2 + 2 \left(t_2(t_1+t_2)+a^2\right)^{\frac{1}{2}}}$$

According to Lemma 4.9 the function w is a rate of convergence of type (2.1) and the corresponding estimate function is

$$s(t_1,t_2) = t_2 - a + \sqrt{t_2(t_1 + t_2) + a^2}.$$

Let us recall what a rate of convergence of type (p.1) means. Consider an interval T of the form $(o,b]$ or $(0,\infty)$ and a positive integer p. Having a function $w:T^p \to T$ we can define the iterates $w^{(n)}$ by taking for $t = (t_1,\ldots,t_p) \in T^p$

$$w^{(0)}(t) = t_p, \quad w^{(n+1)}(t) = w^{(n)}(t_2,\ldots,t_p,w(t)), \quad n = 0,1,\ldots .$$

We have obviously

$$w^{(1)}(t) = w(t), \quad w^{(2)}(t) = w(t_2,\ldots,t_p,w(t)) \text{ etc.}$$

According to Definition 3.2 the function w is called a rate of convergence of type (p.1) if the series $s(t) \sum_{n=0}^{\infty} w^{(n)}(t)$ is convergent for all $t \in T^p$. A rate of convergence of type (p.1) will also be called a p-dimensional rate of convergence.

Now we can state the following

6.9 Definition *Let* $(x_n)_{n \geq 1-p}$ *be a sequence of points of a complete metric space* (X,d) *having the property (14). Let* T *be an interval satisfying the condition (24)*

$$(0, \sup \{d(x_n, x_{n+1}); \ n \geq 1-p\}] \subset T \subset (0, \infty)$$

and let $w: T^p \to T$ *be a rate of convergence of type (p.1). Then*

(i) *We say that* w *is a p-dimensional rate of convergence of the sequence* $(x_n)_{n \geq 1-p}$ *if*

$$d(x_{n+1}, x_n) \leq w(d(x_{n-p+1}, x_{n-p}), \ldots, d(x_n, x_{n-1})), \ n = 1, 2, \ldots . \qquad (30)$$

(ii) *We say that* w *is the exact p-dimensional rate of convergence of the sequence* $(x_n)_{n \geq 1-p}$ *if*

$$d(x_{n+1}, x_n) = w(d(x_{n-p+1}, x_{n-p}), \ldots, d(x_n, x_{n-1})), \ n = 1, 2, \ldots . \qquad (31)$$

Considering nondecreasing p-dimensional rates of convergence - that is rates of convergence of type (p.1) for which $t_1 \leq t_1', \ldots, t_p \leq t_p'$ implies

$$w(t_1, \ldots, t_p) \leq w(t_1', \ldots, t_p')$$

we obtain the following generalization of Proposition 6.8:

6.10 Proposition *Let* $(x_n)_{n \geq 1-p}$ *be a sequence of points of a complete metric space* (X,d). *If this sequence has a nondecreasing p-dimensional rate of convergence* w, *then it converges to a point* $x^* \in X$ *and*

$$d(x_n, x^*) \leq s_1(d(x_{n-p+1}, x_{n-p}), \ldots, d(x_n, x_{n-1}))$$

$$\leq s(\tilde{w}^{(n)}(d(x_{-p+2}, x_{n-p+1}), \ldots, d(x_1, x_0))),$$

where the functions s, s_1 *and* $\tilde{w}$ *are given for any* $t = (t_1, \ldots, t_p) \in T^p$ *by the relations*

$$s(t) = \sum_{n=0}^{\infty} w^{(n)}(t),$$

$$s_1(t) = s(t) - t_p,$$

$$\tilde{w}(t) = (t_2,\ldots,t_p,w(t)).$$

If we have a mapping $F:X^p \to X$ and a p-dimensional rate of convergence w such that

$$d(F(y_2,\ldots,y_p,F_y),Fy) \leqq w(d(y_2,y_1),\ldots,d(y_p,y_{p-1}),d(F_y,y_p))$$

for all $y = (y_1,\ldots,y_p) \in X^p$ then w is obviously a p-dimensional rate of convergence of any sequence $(x_n)_{n \geq -p+1}$ produced by (29).

We have seen that if a sequence $(x_n)_{n \geq 0}$ has a (one-dimensional) rate of convergence w satisfying condition (27) then the number r $(r > 1)$ is a Q-order of convergence of this sequence. An analogous result for p-dimensional rates of convergence can only be given if a more general notion of order of convergence is considered. Such a notion will be introduced in the next paragraph.

R-orders of convergence

In order to unify some formulae we shall use, in what follows, the notation

$$e_n(r) = \begin{cases} n & \text{if } r = 1 \\ r^n & \text{if } r > 1 \end{cases} \qquad n = 0,1,2,\ldots . \tag{32}$$

6.11 __Definition__ *Let* $(x_n)_{n \geq 0}$ *be the sequence of points of a metric space* (X,d) *converging to a point* $x^* \in X$ *and let r be a real number greater than or equal to one.*

(i) *We say that r is an R-order of convergence of the sequence* $(x_n)_{n \geq 0}$ *if there are two constants* $b \in (0,1)$ *and* $(0,\infty)$ *such that*

$$d(x_n,x^*) \leqq B \, b^{e_n(r)}, \qquad n = 0,1,2,\ldots \tag{33}$$

(ii) *We say that r is the exact R-order of convergence of the sequence* $(x_n)_{n \geq 0}$ *if there are constants* $a,b \in (0,1)$, $A,B \in (0,)$ *such that*

$$A \, a^{e_n(r)} \leqq d(x_n,x^*) \leqq B \, b^{e_n(r)}. \tag{34}$$

Let us denote by $R(x_n)$ the set of all R-orders of convergence of the sequence $(x_n)_{n \geq 0}$. It is easy to see that $r \in R(x_n)$ implies $[1,r] \subset R(x_n)$. Denoting

$$\rho^* = \sup_n R(x_n)$$

we have obviously

$$[1,\rho^*) \subset R(x_n) \subset [1,\rho^*].$$

If a sequence $(x_n)_{n\geq 0}$ has an exact R-order of convergence then this order equals ρ^* and $R(x_n) = [1,\rho^*]$. As a consequence it follows that the exact R-order of convergence of a sequence is unique. By analogy with Defintion 6.2 we have

6.12 Definition *Let* $(x_n)_{n\geq 0}$ *be a convergent sequence in a metric space and let* $R(x_n)$ *be the set of all R-orders of convergence of this sequence. Then the number* $\rho^* = \sup_n R(x_n)$ *will be called the weak R-order of convergence of the sequence* $(x_n)_{n\geq 0}$.

We shall show that the notion defined above is intimately related to the asymptotic behaviour of the sequence $(|\log d(x_n,x^*)|^{1/n})_{n\geq 1}$. Before giving this result let us state a lemma which shows that the notions introduced in Definition 6.11 are entirely of an asymptotic nature:

6.13 Lemma *Let* $(x_n)_{n\geq 0}$ *be a convergent sequence of a metric space* (X,d) *with limit* x^*. *Let M and N be two non-negative integers and let r be a real number greater than or equal to one.*

(i) r is an R-order of convergence of $(x_n)_{n\geq 0}$ *if and only if there exist* $b_1 \in (0,1)$ *and* $B_1 \in (0,\infty)$ *such that:*

$$d(x_{N+k},x^*) \leq B_1 b_1^{e_{M+k}(r)} \qquad k = 0,1,2,\ldots \tag{35}$$

(ii) r is the exact R-order of convergence of $(x_n)_{n\geq 0}$ *if and only if there exist* $a_1,b_1 \in (0,1)$ *and* $A_1,B_1 \in (0,\infty)$ *such that:*

$$A_1 a_1^{e_{M+k}(r)} \leq d(x_{N+k},x^*) \leq B_1 b_1^{e_{M+k}(r)}, \qquad k = 0,1,2,\ldots \ . \tag{36}$$

Proof The proof of this lemma being elementary, we sketch only the proof of the sufficiency of condition (35). If this condition is satisfied then, writing

$$B_2 = B_1 b_1^{M-N} \text{ and } b = b_1 \text{ if } r = 1$$

or

$$B_2 = B_1 \text{ and } b = b_1^{r^{M-N}} \text{ if } r > 1,$$

it follows that

$$d(x_{N+k}, x^*) \leq B_2 \, b^{e_{N+k}(r)}, \quad k = 0,1,2,\ldots .$$

Finally, setting

$$B = \max \{B_2, \max \{b^{-e_n(r)} d(x_n, x^*); \ 0 \leq n < N\}\}$$

we obtain relation (33). $\qquad \square$

Lemma 6.13 will be used in the proof of the following result:

6.14 Proposition *If the set $R(x_n)$ is nonvoid then*

$$\sup R(x_n) = \lim \inf |\log d(x_n, x^*)|^{1/n}.$$

Proof If $r \in R(x_n)$ then there exist $b \in (0,1)$ and $B \in (0,\infty)$ such that

$$d(x_n, x^*) \leq B \, b^{e_n(r)} \quad n = 0,1,2,\ldots .$$

As $\lim\limits_{n \to \infty} e_n(r) = \infty$ and $\lim\limits_{n \to \infty} d(x_n, x^*) = 0$ it follows that there exists an $N_1 \in \mathbb{N}$ such that

$$e_n(r) \geq 2|\log B| \ / \ |\log b| \text{ and } d(x_n, x^*) < 1 \text{ for } n \geq N_1.$$

From the above inequalities we deduce that

$$|\log d(x_n, x^*)|^{1/n} \geq (\tfrac{1}{2}|\log b|)^{1/n}(e_n(r))^{1/n} \text{ for } n \geq N_1$$

and because $\lim\limits_{n \to \infty} (e_n(r))^{1/n} = r$ it follows that

$$\lim \inf |\log (d(x_n, x^*))|^{1/n} \geq r.$$

Since r was arbitrarily chosen in $R(x_n)$ this shows that

$$\lim \inf |\log d(x_n, x^*)|^{1/n} \geq \rho^* = \sup R(x_n).$$

Suppose now that

$$\lim \inf \left| \log d(x_n, x^*) \right|^{1/n} > \rho^*.$$

It follows that there exists a number $s > \rho^*$ and an integer $N \geq N_1$ such that

$$\left| \log d(x_n, x^*) \right|^{1/n} \geq s \text{ for } n \geq N.$$

Hence

$$d(x_n, x^*) \leq \left(\frac{1}{e} \right)^{s^n} \text{ for } n \geq N. \tag{37}$$

According to Lemma 6.13 this implies $s \in R(x_n)$ contradicting the fact that $\rho^* = \sup R(x_n)$. □

Suppose now that $(x_n)_{n \geq 0}$ is a sequence for which $R(x_n)$ is nonvoid. The proposition just proved implies then

$$\lim \inf \left| \log d(x_n, x^*) \right|^{1/n} \geq 1. \tag{38}$$

It turns out that this inequality holds even for convergent sequences having no R-order of convergence. Indeed let $(x_n)_{n \geq 0}$ be a sequence converging to x^* and suppose that

$$\lim \inf \left| \log d(x_n, x^*) \right|^{1/n} < 1.$$

This means that there exists a number s and a subsequence $(x_{n_k})_{k \geq 0}$ of $(x_n)_{n \geq 0}$ such that

$$\left| \log d(x_{n_k}, x^*) \right|^{1/n_k} \leq s < 1.$$

From the above inequality we deduce that

$$d(x_{n_k}, x^*) \geq \left(\frac{1}{e} \right)^{s^{n_k}}$$

and letting k tend to infinity we obtain the contradiction $0 = \lim_{k \to \infty} d(x_{n_k}, x^*) \geq 1$.

It is interesting to note that if inequality (38) is strict then the set $R(x_n)$ is nonvoid. More precisely we have

$$[1, \lim \inf \left| \log d(x_n, x^*) \right|^{1/n}) \subset R(x_n).$$

To see this we observe that if

$$s \in (1, \lim \inf |\log d(x_n, x^*)|^{1/n}$$

then there exists a positive integer N for which (37) holds. But we have seen in the proof of Proposition 6.14 that this implies $s \in R(x_n)$.

In the following we shall give sufficient conditions under which a number $r \in [1, \infty)$ is an R-order of convergence of a sequence $(x_n)_{n \geq 0}$. The cases $r > 1$ and $r = 1$ will be investigated separately. By analogy with the terminology introduced in the preceding paragraph we shall say that the sequence $(x_n)_{n \geq 0}$ converges at least R-linearly if 1 is an R-order of convergence of $(x_n)_{n \geq 0}$ and we say that it converges (exactly) R-linearly if 1 is the exact R-order of convergence of $(x_n)_{n \geq 0}$.

6.15 _Lemma_ _If_ $s_1, s_2, \ldots, s_p$ _are_ p _non-negative real numbers satisfying the_ _condition_

$$\sum_{j=1}^{p} s_j > 1 \tag{39}$$

then the equation

$$x^p - \sum_{j=1}^{p} s_j x^{p-j} = 0 \tag{40}$$

has exactly one root belonging to the interval $(1, \infty)$.

The proof of the above lemma is elementary and will be omitted (see for example [64]).

6.16 _Theorem_ _Let_ $(x_n)_{n \geq 0}$ _be a sequence of points of a metric space_ (X, d) _converging to a point_ $x^* \in X$. _Let_ $s_1, s_2, \ldots, s_p$ _be_ p _non-negative real_ _numbers satisfying condition (39) and denote by_ r _the unique root of the_ _equation (40) from the interval_ $(1, \infty)$.

(i) _If there exist_ $C \in (0, \infty)$ _and_ $N \in \mathbb{Z}_+$ _such that_

$$d(x_n, x^*) \quad C \prod_{j=1}^{p} d(x_{n-j}, x^*)^{s_j} \text{ for } n \geq N+p, \tag{41}$$

then $r \in R(x_n)$.

(ii) _If_ $r \in R(x_n)$ _and if there exist_ $D \in (0, \infty)$ _and_ $N \in \mathbb{Z}_+$ _such that_

$$d(x_n,x^*) \geq D \prod_{j=1}^{p} d(x_{n-j},x^*)^{s_j} \quad \textit{for } n \geq N+p, \tag{42}$$

then r is the exact R-order of convergence of $(x_n)_{n\geq 0}$.

<u>Proof</u> We shall prove only point (i), because point (ii) can be proved by analogous methods. Let us consider the sequence $(u_n)_{n\geq 1}$ defined as follows

$$u_i = 0 \text{ for } i = 1,2,\ldots,p, \tag{43}$$

$$u_n = 1 + \sum_{k=1}^{p} s_k\, u_{n-k} \text{ for } n \geq p+1. \tag{44}$$

Let us also consider the sequence $(v_n)_{n\geq 0}$ with $v_n = r^n$. It is easily seen that this is the unique sequence verifying the conditions:

$$v_i = r^i,\ i = 1,2,\ldots,p, \tag{45}$$

$$v_n = \sum_{k=1}^{p} s_k\, v_{n-k},\ n \geq p + 1. \tag{46}$$

Let us prove now that

$$v_n > (s-1)\, u_n + 1 \tag{47}$$

for all $n \in \mathbb{N}$ where s stands for $s = \sum_{i=1}^{p} s_i$.

For $n = 1,2,\ldots,p$ (47) follows immediately from (43) and (45). Let us suppose that (47) holds for $n = 1,2,\ldots,k-1$, with $k \geq p + 1$. According to (39), (44) and (46) we have

$$v_k = \sum_{j=1}^{p} s_j\, v_{k-j} > \sum_{j=1}^{p} s_j((s-1)u_{k-j} + 1) = (s-1)u_k + 1.$$

Thus (47) holds for all $n \in \mathbb{N}$.

Because $\lim_{n\to\infty} d(x_n,x^*) = 0$ there is an $N_1 \geq N$ such that

$$d(x_{N_1+i},x^*) \leq b_1^{r^i},\ i = 1,2,\ldots,p \tag{48}$$

where b_1 is a number from $(0,1)$ having the property that $b_1 c^{\frac{1}{s-1}} < 1$.

From (41), (44), (46) and (48) we deduce by induction that

$$d(x_{N_1+k}, x^*) \leq b_1^{v_k} \cdot c^{u_k} \text{ for } k = 1, 2, 3, \ldots .$$

We may obviously suppose that $C > 1$ and in this case using (47) we obtain

$$b_1^{v_k} c^{u_k} \leq C^{\frac{1}{1-s}} \left[b_1 c^{\frac{1}{s-1}} \right]^{v_k}.$$

Denoting $b = b_1 c^{\frac{1}{s-1}}$ and $B = C^{\frac{1}{1-s}}$ it follows from the above relation and from (49) that

$$d(x_{N_1+k}, x^*) \leq B \, b^{r^k}, \quad k = 1, 2, \ldots \tag{49}$$

and according to Lemma 6.13 this means that $r \in R(x_n)$. □

If the non-negative numbers $s_1, s_2, \ldots, s_p$ satisfy the condition

$$\sum_{j=1}^{p} s_j = 1 \tag{50}$$

then $x = 1$ verifies (40) and this equation has no other root in the interval $[1, \infty)$. In this case we have the following result:

6.17 Theorem *Let $(x_n)_{n \geq 0}$ be a sequence in a metric space (X, d) which converges to a point $x^* \in X$, and let $s_1, \ldots, s_p$ be p non-negative real numbers satisfying equality (50).*

(i) If there exist a constant $C \in (0,1)$ and a non-negative integer N such that condition (41) is satisfied then the sequence $(x_n)_{n \geq 0}$ converges at least R-linearly.

(ii) If the sequence $(x_n)_{n \geq 0}$ converges at least R-linearly and if there exist $D \in (0,1)$ and $M \in \mathbb{Z}_+$ such that condition (42) is satisfied then $(x_n)_{n \geq 0}$ converges exactly R-linearly.

The proof of the above theorem is left as an exercise. (The interested reader should consult, for example, [64]).

Taking $p = 1$ in Theorems 6.16 and 6.17 we obtain the following

6.18 Corollary *If $(x_n)_{n \geq 0}$ is a convergent sequence of a metric space then $Q(x_n) \subset R(x_n)$. Moreover if a number $r \geq 1$ is the exact Q-order of convergence of the sequence $(x_n)_{n \geq 0}$ then r is also the exact R-order of convergence*

of this sequence.

We note that the converse of the above corollary is false. Indeed consider, for any $r \in [1,\infty)$, the sequence of real numbers given by

$$x_n = \begin{cases} a^{e_n(r)} & \text{for n even} \\ b^{e_n(r)} & \text{for n odd} \end{cases}$$

where $0 < a < b < 1$. It is easy to see that r is the exact R-order of convergence of this sequence but $r \notin Q(x_n)$.

6.19 <u>Remark</u> It follows from Theorem 6.16 that if a sequence $(x_n)_{n \geq 0}$ has a p-dimensional rate of convergence $w: T^p \to T$ such that

$$\limsup_{t_i \downarrow 0; i=1,\ldots,p} \frac{w(t_1,\ldots,t_p)}{t_1^{s_1} \ldots t_p^{s_p}} < \infty \, , \tag{51}$$

where $s_1,\ldots,s_p$ are p non-negative real numbers satisfying condition (39) then the unique solution $r \in (1,\infty)$ of the equation (40) is an R-order of convergence of the sequence $(d(x_n,x_{n+1}))_{n \geq 0}$. If $(x_n)_{n \geq 0}$ has an exact p-dimensional rate of convergence satisfying (51) as well as

$$\liminf_{t_i \downarrow 0; i=1,\ldots,p} \frac{w(t_1,\ldots,t_p)}{t_1^{s_1} \ldots t_p^{s_p}} > 0, \tag{52}$$

then r is the exact R-order of convergence of the sequence $(d(x_n,x_{n+1}))_{n \geq 0}$.

In the preceding paragraph we have seen that a number $r > 1$ was a Q-order of convergence of a sequence $(x_n)_{n \geq 0}$ if and only if it was a Q-order of convergence of the sequence $(d(x_n,x_{n+1}))_{n \geq 0}$. In case of R-orders of convergence a corresponding statement holds for $r \geq 1$.

6.20 <u>Proposition</u> *If* $(x_n)_{n \geq 0}$ *is a convergent sequence of a metric space then* $R(x_n) = R(d(x_n,x_{n+1}))$.

<u>Proof</u> The inclusion $R(x_n) \subset R(d(x_n,x_{n+1}))$ follows immediately by applying the triangle inequality

$$d(x_n,x_{n+1}) \leq d(x_n,x^*) + d(x_{n+1},x^*).$$

If $r \in R(d(x_n, x_{n+1}))$ then there are $b \in (0,1)$ and $B \in (0,\infty)$ such that

$$d(x_n, x_{n+1}) \le B \, b^{e_n(r)}, \quad n = 0,1,2,\dots .$$

Hence

$$d(x_n, x^*) \le \sum_{k=0}^{\infty} d(x_{n+k}, x_{n+k+1}) \le B \sum_{k=0}^{\infty} b^{e_{n+k}(r)}. \tag{53}$$

If $r = 1$ then we have

$$\sum_{k=0}^{\infty} b^{e_{n+k}(r)} = b^n \sum_{k=0}^{\infty} b^k = \frac{b^n}{1-b}, \tag{54}$$

and if $r > 1$ then $e_{n+k}(r) = r^n r^k \ge r^n(1 + (r-1)k)$ and we may write

$$\sum_{k=0}^{\infty} b^{e_{n+k}(r)} \le b^{r^n} \sum_{k=0}^{\infty} [b^{(r-1)r^n}]^k = \frac{b^{r^n}}{1-b^{(r-1)rn}} \le \frac{b^{r^n}}{1-b^{r-1}}. \tag{55}$$

From (53) – (55) it follows that $r \in R(x_n)$. The proof is complete. $\quad\square$

If the sequences $(x_n)_{n \ge 0}$ and $(d(x_n, x_{n+1}))_{n \ge 0}$ have exact R-orders of convergence then, according to the above proposition they must have the same exact R-order of convergence. However (in contrast to point (ii) of Corollary 6.6) the fact that the sequence $(x_n)_{n \ge 0}$ has an exact R-order of convergence does not imply that the sequence $(d(x_n, x_{n+1}))_{n \ge 0}$ has an exact R-order of convergence and vice versa. Indeed let c, r, s be three real numbers such that $0 < c < 1 < r < s$ and set

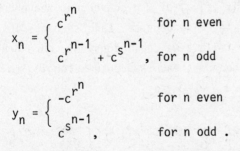

$$x_n = \begin{cases} c^{r^n} & \text{for } n \text{ even} \\ c^{r^{n-1}} + c^{s^{n-1}}, & \text{for } n \text{ odd} \end{cases}$$

$$y_n = \begin{cases} -c^{r^n} & \text{for } n \text{ even} \\ c^{s^{n-1}}, & \text{for } n \text{ odd} . \end{cases}$$

It is easy to see that r is the exact R-order of convergence of the sequences $(x_n)_{n \ge 0}$ and $(d(y_n, y_{n+1}))_{n \ge 0}$, while the sequences $(d(x_n, x_{n+1}))_{n \ge 0}$ and $(y_n)_{n \ge 0}$ have no exact R-order of convergence. Nevertheless we have the following result:

6.21 Proposition *Let* $(x_n)_{n\geq 0}$ *be a sequence of points of a metric space* (X,d) *converging to a point* $x^* \in X$ *such that*

$$d(x_{n+1}, x^*) \leq C\, d(x_n, x^*), \quad n = 0, 1, 2, \ldots \tag{56}$$

where C *is a positive constant. If a number* $r \in [1, \infty)$ *is the exact R-order of convergence of the sequence* $(d(x_n, x_{n+1}))_{n\geq 0}$ *then* r *is also the exact R-order of convergence of* $(x_n)_{n\geq 0}$.

Proof   The result follows immediately using Proposition 6.20 and observing that:

$$(1 + C)\, d(x_n, x^*) \geq d(x_n, x^*) + d(x_{n+1}, x^*) \geq d(x_n, x_{n+1}). \qquad \square$$

In some cases condition (56) can be verified from theoretical considerations without knowing the root x^*. Thus the above proposition allows us to compute the exact R-order of convergence of a sequence $(x_n)_{n\geq 0}$ from the behaviour of the sequence $(d(x_n, x_{n+1}))_{n\geq 0}$. This can be useful if we know for example the fact that $(x_n)_{n\geq 0}$ has an exact p-dimensional rate of convergence satisfying conditions (51) and (52).

Orders of convergence and efficiency

Until now we have studied the convergence of individual sequences. However, dealing with iterative procedures we are led to the consideration of classes of convergent sequences. Indeed an iterative procedure of type (p.m) for a class C will associate with each pair (f, v_0) of C m convergent sequences. Let us denote by S the class of all those sequences. Each sequence in S belongs to a Banach space. In what follows we shall consider, more generally, classes S composed of convergent sequences, each sequence belonging to a metric space.

6.22 Definition *Let* S *be a class of convergent sequences and let* r *be a real number greater than or equal to one.*

(i) *We say that* r *is a Q-order (resp. R-order) of convergence of the class* S *if* r *is a Q-order (resp. R-order) of convergence of every sequence of* S.

(ii) *We say that* r *is the exact Q-order (resp. R-order) of the class* S *if* r

is a Q-order (resp. R-order) of any sequence of S and if there exist a sequence in S which has the exact Q-order (resp. R-order) of convergence equal to r.

6.23 <u>Definition</u> *If S is a class of convergent sequences then the numbers*

$$\inf_{(x_n)\in S} \sup_n Q(x_n), \quad \inf_{(x_n)\in S} \sup_n R(x_n)$$

are called respectively the weak Q-order of convergence of the class S and the weak R-order of convergence of the class S.

We note that if r is a Q-order (resp. the exact Q-order) of convergence of the class S then r is also an R-order (resp. the exact R-order) of convergence of this class. If the class S is composed of only one sequence then the definitions given above coincide with the definitions introduced in the preceding sections.

In what follows, when speaking about the order of convergence of an iterative procedure we shall always mean the order of convergence of the class S of sequences produced by it. Thus if a number $r > 1$ is an R-order of an iterative procedure and if $(x_n)_{n\geq 0}$ is a sequence produced by this iterative procedure, with limit x^*, then there are $B \in (0,\infty)$ and $b \in (0,1)$ such that

$$d(x_n,x^*) \leq B\, b^{r^n}, \quad n = 0,1,2,\ldots .$$

It follows that, given an $\varepsilon > 0$, we have $d(x_n,x^*) \leq \varepsilon$ for all integers n greater than

$$N(\varepsilon) = \frac{1}{\log_2 r} \log_2 \left\lceil \frac{\log_2 \varepsilon - \log_2 B}{\log_2 b} \right\rceil .$$

Hence the number of steps required to reach a desired precision is proportional to $1/\log_2 r$. This shows that a good measure for the efficiency of an iterative procedure is the ratio

$$\text{eff} = \frac{\log_2 r}{c} \tag{57}$$

where r is the exact R-order of convergence and c the cost of a step of the iterative procedure. Generally c is a constant proportional to the computer time required for performing a step of the iterative procedure at a given computer. In other words c is proportional to the number of elementary arithmetic operations (e.a.o.) to be performed at each step of the iterative procedure (an e.a.o. is a standard operation by means of which one can express the execution time of the other operations at a given computer; for example: 1 addition = 1 e.a.o., 1 multiplication = 1.8 e.a.o., 1 division = 2.3 e.a.o.).

Applications

In this paragraph we shall discuss the order of convergence and the efficiency of the iterative procedures studied in Sections 4 and 5. We shall refer to those iterative procedures using the number of the iterative algorithm defining them. Thus by "the iterative procedure (4.2)" we mean the simplified secant method for the class $C(h_0, r_0, q_0)$ and by "the iterative procedure (5.1)" we mean Newton's method for the class $C'(k_0, r_0)$.

For computing the efficiency we need only the R-order of convergence but whenever possible we shall give the Q-order of convergence.

In the statement of the following propositions we shall (tacitly) assume that the inequalities (4.7) and (5.6) appearing in the definition of the class $C(h_0, r_0, q_0)$ and $C'(k_0, r_0)$ are strict. This means that the constants defined by (4.12) and (5.8) do not vanish.

6.24 **Proposition** *The exact Q-order of convergence of the iterative procedure (4.2) equals one.*

Proof Let $(x_n)_{n \geq 0}$ be the sequence corresponding to a given triplet $(f, x_0, x_1) \in C(h_0, r_0, q_0)$. Using (4.9) we obtain

$$|x_{n+1} - x^*| = |x_n - x^* - T_0(f(x_n) - f(x^*))|$$

$$(58)$$

$$= |T_0(\delta f(x_{-1}, x_0)(x_n - x^*) - f(x_n) + f(x^*))|$$

$$\leq h_0(|x_n - x_0| + |x^* - x_0| + |x_0 - x_{-1}|) |x_n - x^*|.$$

According to Theorem 4.3 we have

$$|x_n - x_0| < \mu_0, \quad |x^* - x_0| \leq \mu_0$$

so that

$$|x_{n+1} - x^*| \le h_0(2\mu_0 + q_0) |x_n - x^*| = (1 - 2ah_0) |x_n - x^*|. \tag{59}$$

This shows that 1 is a Q-order of convergence of the iterative procedure (4.2).

If $(x_n)_{n \ge 0}$ is the sequence corresponding to the triplet (f, x_0, x_{-1}) considered in the proof of Proposition 4.4 then $x_{n+1} - a = (1-h_0(x_n+a))(x_n-a)$. Hence 1 is the exact Q-order of convergence of the iterative procedure (5.2). □

6.25 Corollary *The exact Q-order of convergence of the iterative procedure (5.5) is equal to one.*

6.26 Proposition *The exact Q-order of convergence of the iterative procedure (5.1) is equal to two.*

Proof From the proof of Theorem 5.3 it follows that the function $w(t) = \frac{1}{2} t^2 (t^2 + a^2)^{-1/2}$ is the exact rate of convergence of the iterative procedure (5.1). Observing that

$$0 < \frac{1}{2a} = \lim_{t \downarrow 0} \frac{w(t)}{t^2} < \infty$$

we deduce that the exact C-order of convergence of the iterative procedure (5.1) equals two.   □

6.27 Proposition *The exact R-order of convergence of the iterative procedure (4.3) equals $\frac{1}{2}(m + (m^2 + 4)^{1/2})$.*

Proof Consider the rate of convergence of type (2.m) introduced in Lemma 4.12 and set

$$y_n^k = s_k(\tilde{w}^{(n-1)}(q_0, r_0)), \ 0 \le k \le m, \ n = 1, 2, \ldots \ .$$

At the end of Section 4 we have remarked that the estimates (4.31) are attained for the triplet (f, x_0, x_{-1}) considered in Proposition 4.4. It follows that for proving our proposition it is sufficient to show that the sequences $(y_n^j)_{n \ge 0}, \ 1 \le j \le m$ have the exact R-order of convergence $\frac{1}{2}(m + (m^2 + 4)^{1/2})$. One can easily show that these sequences are related by the formula

$$y_{n+1}^{k+1} = \frac{y_{n+1}^k (y_n^m + y_n^{m-1} - y_{n+1}^k)}{y_n^m + y_n^{m-1} + 2a} \,. \tag{60}$$

It follows that

$$0 < y_{n+1}^m < y_{n+1}^{m-1} < \cdots < y_{n+1}^0 = y_n^m. \tag{61}$$

From the above relations we obtain successively:

$$y_{n+1}^{k+1} \le \frac{1}{a} y_{n+1}^k y_n^{m-1} \le \cdots \le (\frac{1}{a})^{k+1} y_n^m (y_n^{m-1})^{k+1}$$

$$\le (\frac{1}{a})^{k+2} (y_n^{m-1})^{k+2} y_{n-1}^{m-1}.$$

Hence $y_{n+1}^{m-1} \le (\frac{1}{a})^m (y_n^{m-1})^m y_{n-1}^{m-1}$ and then according to Theorem 6.16 it follows that $r \in R(y_n^{m-1})$ where r is the positive root of the equation $x^2 - mx - 1 = 0$ i.e.

$$r = \frac{1}{2} (m + (m^2 + 4)^{1/2}).$$

On the other hand from (60) and (61) we obtain

$$y_{n+1}^{k+1} \ge \frac{y_{n+1}^k y_n^{m-1}}{y_0^m + y_0^{m-1} + 2a} = h_0 y_{n+1}^k y_n^{m-1} \ge \cdots \ge h_0^{k+1} y_n^m (y_n^{m-1})^{k+1}$$

$$\ge h_0^{k+2} (y_n^{m-1})^{k+2} y_{n-1}^{m-1}.$$

For k = m-2 this gives $y_{n+1}^{m-1} \ge h_0^m (y_n^{m-1})^m y_{n-1}^{m-1}$ and using again Theorem 6.16 we deduce that r is the exact R-order of convergence of the sequence $(y_n^{m-1})_{n \ge 0}$. The proof of our proposition is now complete, because it follows from (61) that if one of the sequences $(y_n^j)_{n \ge 1}$, $0 \le j \le m$, has an exact R-order of convergences then all these sequences will have the same exact R-order of convergence. □

For m = 1 we obtain the following well known result concerning the rapidity of the secant method:

6.28 <u>Corollary</u> *The iterative procedure (4.1) has the exact R-order of con-vergence* $(1 + \sqrt{5})/2$.

We note that this result can be deduced directly observing that the

116

function w given by (4.19) is the exact rate of convergence of the iterative procedure (4.1) and applying Remark 6.19.

Finally, for Traub's method we have:

6.29 Proposition *The iterative procedure (5.3) has the exact R-order of convergence* $m + 1$.

Proof Considering the notation used in Lemma 5.8 and Theorem 5.9 and denoting $y_n^j = s_j \circ w_m^{(n-1)}(r_0)$ we have:

$$y_1^0 = k_0^{-1} - a, \quad y_{n+1}^{k+1} = \frac{y_{n+1}^k (2y_n^m - y_{n+1}^k)}{2y_n^m + 2a}, \quad 0 \le k \le m-1, \ n \ge 0 \qquad (62)$$

$$0 < y_{n+1}^m < y_{n+1}^{m-1} < \cdots < y_{n+1}^1 < y_{n+1}^0 = y_n^m. \qquad (63)$$

It follows that

$$y_{n+1}^m \le \frac{1}{a} y_{n+1}^{m-1} y_n^m \le (\frac{1}{a})^2 y_{n+1}^{m-2} (y_n^m)^2 \le \cdots \le (\frac{1}{a})^m (y_n^m)^{m+1},$$

from which we see that $m + 1 \in Q(y_n^m) \subset R(y_n^m)$.

On the other hand from (62) and (63) we easily deduce that

$$y_{n+1}^{k+1} > \frac{y_{n+1}^k y_n^m}{2y_1^0 + 2a} > (\frac{k_0}{4}) y_{n+1}^k y_n^m.$$

By iterating this inequality we get

$$y_{n+1}^m > (\frac{k_0}{4})^m (y_n^m)^{m+1}.$$

This shows that $m + 1$ is the exact Q-order (and consequently the exact R-order) of convergence of the sequence $(y_n^m)_{n\ge1}$. From (63) it follows that $m + 1$ is the exact R-order of convergence of any sequence $(y_n^j)_{n\ge1}$ with $1 \le j \le m$. By virtue of Theorem 5.9 and Proposition 5.10 the proof of our proposition is complete. □

6.30 Remark It can be easily proved that $m + 1$ is a Q-order of convergence of any sequence $(x_n^m)_{n\ge1}$ produced by the iterative procedure (5.3). However for the sequences $(x^j)_{n\ge1}$ with $1 \le j \le m-1$ we can only prove that $m + 1$ is an R-order of convergence.

In the sequel we shall discuss the efficiency of the iterative procedures (4.1), (4.3), (5.1), (5.3). The notion of efficiency is related to effective numerical applications so that it makes sense only in the finite dimensional case. For this reason in what follows we shall consider only operators $f:D \subset \mathbb{R}^d \to \mathbb{R}^d$ where d is a given positive integer.

First let us study the secant method (4.1). The consistent approximations of the derivative used by this iterative procedure will be taken to be divided differences of the form (13.10). With this choice let us analyse the computations to be done for performing a step of the iterative procedure (4.1). Because the values $f_1(x_{n-1}),\ldots,f_d(x_{n-1})$ have already been computed at the preceding step it follows that, to perform one step of the iterative procedure (4.1), it is necessary to make d^2 function evaluations. Let us suppose that each function evaluation requires N elementary arithmetic operations (e.a.o.). Let N be the rest of e.a.o. necessary for performing formulae (4.1) (i.e. the e.a.o. necessary for solving a linear system). Then the efficiency of the secant method is

$$\frac{\log_2 (1 + \sqrt{5}) - 1}{N\,d^2 + M} \tag{64}$$

In a similar way it follows that the efficiency of the iterative procedure (4.3) is

$$\frac{\log_2 (m + \sqrt{m^2 + 4}) - 1}{(d + m - 1)\ Nd + Mn} \tag{65}$$

For m = 1 the iterative procedure (4.3) reduces to the secant method (4.1) so that in this case $M_1 = M$, and (65) coincides with (64).

Let us analyse now Newton's method. In case the partial derivatives $\frac{\partial f_i}{\partial x_j}$ have a simple analytical form (for example if the functions f_i are polynomials) then we may consider that the computation of one value of a function $\frac{\partial f_i}{\partial x_j}$ requires also N e.a.o. In this case we need Nd^2 e.a.o. for computing the Jacobian $f'(x_n)$. To this quantity we have to add Nd e.a.o. required for the computation of the vector $f(x_n)$. In case the partial derivatives $\frac{\partial f_i}{\partial f_j}$ do not have a simple analytical expression then the Jacobian will be

118

approximated by taking

$$\frac{\partial f_i}{\partial x_j}(x_1,\ldots,x_d) \cong \frac{1}{h}[f_i(x_1,\ldots,x_j + h,\ldots,x_d) - f_i(x,\ldots,x_j,\ldots,x_d)].$$

Thus we may conclude that in both cases the computation of the Jacobian $f'(x_n)$ and of the vector $f(x_n)$ requires $(d^2 + d)N$ e.a.e. Hence the efficiency of Newton's method is equal to

$$\frac{1}{(d^2 + d)N + M} \tag{66}$$

Analogously, the efficiency of the iterative procedure (5.3) is

$$\frac{\log_2 (m + 1)}{(d^2 + md)N + M_n} \tag{67}$$

If the dimension d is small then N is generally much bigger than M_n. In the following table we have labelled the efficiency of the iterative procedures (4.1), (4.3), (5.1), (5.3) for d = 1,2,3. We have taken N = 1 and M_m = 0. The efficiency of the secant method has been displayed separately. For the iterative procedures (4.3) and (5.3) we have displayed the maximum efficiency and the value of the parameter m for which this maximum is attained.

dimension procedure	d = 1	d = 2	d = 3
(4.1)	0.69425191	0.17356048	0.07713799
(5.1)	0.5	0.16666667	0.08333333
(4.3)	0.69424191 (m=1)	0.21545987 (m=3)	0.11570699 (m=4)
(5.3)	0.52832083 (m=2)	0.2 (m=3)	0.11111111 (m=3)

119

7 An eigenvalue problem

In this section we shall use the method of nondiscrete induction to investigate the spectrum of some bounded linear operators on a Banach space E. We restrict ourselves to the finite-dimensional case.

Suppose the space E is written as a direct sum $E = E_1 \oplus E_2$ where E_1 is one-dimensional. Given $u \in E_2$, $v \in {}^d E_2$, $C \in B(E_2)$ and $a_{11} \in \mathbb{C}$ we write

$$A = \begin{pmatrix} a_{11} & v \\ u & C \end{pmatrix} \tag{1}$$

for the operator $A \in B(E)$ which associates to each $x = (x_1, x_2)^T$ a $y = (y_1, y_2)^T$ defined as follows

$$y_1 = x_1 a_{11} + \langle x_2, v \rangle$$

$$y_2 = x_1 u + C x_2.$$

If the product $a = |u| \, |v|$ vanishes then a_{11} will be obviously an eigenvalue of A. If, in addition, the operator $C - a_{11}$ is invertible then the eigenvalue a_{11} will be simple i.e. the corresponding eigenspace will be one-dimensional. In fact one can easily see that A has a simple eigenvalue if and only if it can be written in the form (1) with $|u| \, |v| = 0$ and $C - a_{11}$ invertible.

In this case the open disk $\{z; |z - a_{11}| < b\}$, where $b = d(C - a_{11})$, will contain no other points of the spectrum of A.

The main purpose of this section is to show that the operator A still has spectral properties of this type, even when the product $a = |u| \, |v|$ does not vanish but is sufficiently small. More precisely we shall prove that if $4a \leq b^2$ then the operator A has a simple eigenvalue λ belonging to the closed disk $\{z; |z-a_{11}| \leq \frac{1}{2} (b - (b^2 - 4a)^{1/2})\}$ and that the intersection of the spectrum of A with the open disk $\{z; |z-a_{11}| < \frac{1}{2} (b + (b^2 - 4a)^{1/2})\}$ contains no points different from λ.

First, let us state the following lemma:

7.1 Lemma *Let λ be a point in the resolvent set of* C. *Then:*

(i) *if λ belongs to the spectrum of* A *then*

$$a_{11} - \lambda + \langle(\lambda-C)^{-1}u,v\rangle = 0; \tag{2}$$

(ii) *if λ satisfies (2) then λ is a simple eigenvalue of* A *and the corresponding eigenspace is equal to* $Q(\lambda)E_1$ *where* $Q(\lambda)$ *is the bounded linear operator given by the matrix:*

$$Q(\lambda) = \begin{pmatrix} 1 & 0 \\ (\lambda-C)^{-1}u & I \end{pmatrix}. \tag{3}$$

<u>Proof</u> Use the identity

$$(A - \lambda)\,Q(\lambda) = \begin{pmatrix} a_{11} - \lambda + \langle(\lambda-C)^{-1}u,v\rangle & v \\ 0 & C-\lambda \end{pmatrix}$$

and the fact that $Q(\lambda)$ is nonsingular. □

Using the fact that $C - a_{11} - z$ is invertible for all $|z| < b$ we obtain the following.

7.2 Corollary *The intersection of the spectrum of* A *with the open disk* $\{z; |z - a_{11}| < b\}$ *is included in the point spectrum of* A.

Now let us introduce a rate of convergence to be used in the proof of our main theorem.

7.3 Lemma *Let* a *and* b *be two numbers satisfying the inequalities* $a \geq 0$, $b > 0$, $4a \leq b^2$ *and let us consider the function*

$$u(r) = ((b + r)^2 - 4a)^{1/2}. \tag{4}$$

Then the function

$$w(r) = r\,\frac{b + r - u(r)}{b - r + u(r)} \tag{5}$$

is a rate of convergence on the interval $T = \{t; t > 0\}$ *and the corresponding estimate function is given by*

$$s(r) = \frac{1}{2} (r + u(r)) - \frac{1}{2} (b^2 - 4a)^{1/2}. \tag{6}$$

<u>Proof</u> Verify the identity $s(w(r)) + r = s(r)$ and use Proposition 1.2. □

We can state now the main theorem:

7.4 <u>Theorem</u> *Let E be a Banach space written as a direct sum $E = E_1 \oplus E_2$
where E_1 is one-dimensional. Let a_{11} be a complex number and let $u \in E_2$,
$v \in {}^d E_2$, $C \in B(E_2)$ be given. Suppose the linear operator $B = C - a_{11}$ is
invertible and set $b = d(B)$, $a = |u| \, |v|$. If $4a \leq b^2$ then:*

(i) *the linear operator $A \in B(E)$, given by (1), has a simple eigenvalue λ
such that*

$$|a_{11} - \lambda| \leq \frac{1}{2} (b - (b^2 - 4a)^{1/2}); \tag{7}$$

(ii) *the eigenspace corresponding to λ is $Q(\lambda)E_1$ where $Q(\lambda)$ is given by (3);*

(iii) *λ is the limit of the sequence (λ_n) generated via the algorithm*

$$\lambda_0 = a_{11}, \ \lambda_{n+1} = a_{11} + \langle(\lambda_n - C)^{-1}u, v\rangle, \ n = 0,1,\ldots \tag{8}$$

(iv) *the following inequalities are satisfied*

$$|\lambda_n - \lambda| \leq s(w^{(n)}(|\langle B^{-1}u, v\rangle|)) \leq s(w^{(n)}(a/b)), \ n = 0,1,2,\ldots \tag{9}$$

$$|\lambda_n - \lambda| \leq s(|\lambda_n - \lambda_{n-1}|) - |\lambda_n - \lambda_{n-1}|, \ n = 1,2,\ldots \tag{10}$$

where w and s are the functions given in Lemma 7.3.

<u>Proof</u> The proof is based on Proposition 1.9. First let us observe that
with the substitution $z_n = \lambda_n - a_{11}$ the iterative algorithm (8) can be rewritten under the form

$$z_0 = 0, \ z_{n+1} = G(z_n), \ n = 0,1,\ldots \tag{8'}$$

where $G(z) = \langle(z - B)^{-1}u, v\rangle$. Evidently, the domain D of G is composed of
those z for which $z - B$ is invertible. We shall attach to this iterative
algorithm the rate of convergence w obtained in Lemma 7.3 and the approximate

set

$$Z(r) = \{z; d(z - B) \geq h(r), |z - G(z)| \leq r\}$$

where

$$h(r) = \frac{1}{2} (b + r + u(r)) = s(r) + \frac{1}{2} (b + (b^2 - 4a)^{1/2}).$$

Let us note that the condition $d(z - B) \geq h(r)$ implies the invertibility of $z - B$ because $h(r) > \frac{b}{2} > 0$ for all $r > 0$. Observing that

$$h(|<B^{-1}u,v>|) \leq h(a/b) = \frac{1}{2b} (b^2 + a + ((b^2 + a)^2 - 4ab^2)^{1/2}) = b$$

and

$$|G(0)| = |<B^{-1}u,v>|$$

it follows that the inclusion $z_0 \in Z(r_0)$ will be satisfied for $z_0 = 0$ and $r_0 = |<B^{-1}u,v>|$.

We intend to prove now that the inclusion $z \in Z(r)$ implies $z' = G(z) \in Z(w(r))$. If $z \in Z(r)$ then we have

$$d(z' - B) \geq d(z - B) - |z - z'| \geq h(r) - r.$$

The function h differs from the estimate function s only by an additive constant so that it will satisfy the relation $h(r) - r = h(w(r))$. Thus $z \in Z(r)$ implies $d(z' - B) \geq h(w(r))$. Now we may write

$$|z' - G(z')| = |z' - <(z'-B)^{-1}u,v>| = |<((z-B)^{-1} - (z'-B)^{-1})u,v>|$$

$$\leq |(z - B)^{-1}| |(z'-B)^{-1}| |u| |v| |z - z'| \leq \frac{ar}{h(r)(h(r)-r)} .$$

It is easy to check that $ar = h(r) (h(r) - r) w(r)$, so that $|z'-G(z')| \leq w(r)$. Thus conditions (1.7) - (1.9) of Proposition 1.9 are satisfied. Consequently the sequence z_n generated by (8') will converge to a point z* and the following relations will be satisfied:

$$z_n \in Z(w^{(n)}(r_0)), \quad n = 0,1,2,\ldots \tag{11}$$

$$|z_n - z^*| \leq s(w^{(n)}(r_0)) \quad n = 0,1,2,\ldots . \tag{12}$$

Letting n tend to infinity in (8') we get the relation $z^* = \langle (z^*-B)^{-1}u, v\rangle$.
According to Lemma 7.1 this means that $\lambda = z^* + a_{11}$ is a simple eigenvalue
of A and that its corresponding eigenspace is $Q(\lambda)E_1$. Substituting
$\lambda_n = z_n + a_{11}$ in (12) we obtain (9).

For n = 0 this reduces to (7). According to point 3. of Proposition 1.9
the estimates (10) will also be satisfied provided that

$$z_{n-1} \in Z(|z_n - z_{n-1}|) . \tag{13}$$

From (11) it follows that $|z_n - z_{n-1}| \leq w^{(n-1)}(r_0)$ and $d(z_{n-1}-B) \geq h(w^{(n-1)}(r_0))$.
Using the fact that h is increasing we obtain $d(z_{n-1}-B) \geq h(|z_n-z_{n-1}|)$ and
hence (13). □

The following proposition shows that the estimates given in Theorem 7.4
are sharp.

7.5 Proposition *Let $a \geq 0$, $b > 0$ be two numbers satisfying the inequality
$4a \leq b^2$. Then the hypotheses of Theorem 7.4 are verified for $E_1 = E_2 = \mathbb{C}$,
$a_{11} = 0$, $u = -a^{1/2}$, $v = a^{1/2}$, $c = b$ and the estimates (9) and (10) are
attained for all n.*

Proof The 2×2 matrix $A = \begin{pmatrix} 0 & a^{1/2} \\ -a^{1/2} & b \end{pmatrix}$ has the eigenvalues

$$p_1 = \frac{1}{2}(b - (b^2 - 4a)^{1/2}) \text{ and } p_2 = \frac{1}{2}(b + (b^2 - 4a)^{1/2}).$$

Only p_1 satisfies (7) so that the sequence (λ_n) generated by (8) will converge to this eigenvalue. For n = 0 (9) reduces to

$$|\frac{1}{2}(b - (b^2 - 4a)^{1/2})| \leq s(a/b) = \frac{1}{2}(b - (b^2 - 4a)^{1/2}).$$

Hence according to Proposition 1.11 (9) will be verified with equality for
all n. The rate of convergence w being nondecreasing it follows that (10)
will be also attained for all n (see Proposition 1.10). □

From the above proposition it follows that

$$s(w^n(a/b)) = \frac{1}{2}(b - (b^2 - 4a)^{1/2}) - s_n, w^n(a/b) = s_{n+1} - s_n \tag{14}$$

where s_n is the sequence generated by the following algorithm

$$s_0 = 0, \quad s_{n+1} = a/(b - s_n), \quad n = 0,1,2,\ldots . \tag{15}$$

It turns out that it is possible to find an explicit form for the terms of this sequence. A way of doing it is given in Appendix C. In this section we shall content ourselves in stating the result which can easily be verified by induction.

7.6 Proposition *Let* $a \geq 0$ *and* $b > 0$ *be two numbers satisfying the inequality* $4a \leq b^2$ *and let* (s_n) *be the sequence generated by (15);*

(i) *if* $4a < b^2$ *then*

$$s_n = p \, \frac{1-q^n}{1 - q^{n+1}} \quad n = 0,1,2,\dots \tag{16}$$

where

$$p = \frac{1}{2} \, (b - (b^2 - 4a)^{1/2}), \quad q = 2p/(b + (b^2 - 4a)^{1/2}); \tag{17}$$

(ii) *if* $4a = b^2$ *then*

$$s_n = \frac{n}{2n + 2}b \quad n = 0,1,2,\dots \, . \tag{18}$$

Using (14) and the result of the above proposition we can give the apriori estimates (9) an explicit form.

7.7 Corollary *Let* $a \geq 0$ *and* $b > 0$ *be two numbers satisfying the inequality* $4a \leq b^2$ *and let* w *and* s *be the functions given in Lemma 7.3;*

(i) *if* $4a < b^2$ *then*

$$s(w^n(a/b)) = \frac{p(1 - q)}{1 - q^{n+1}} \, q^n, \quad n = 0,1,2,\dots \tag{19}$$

where p *and* q *are the numbers defined by (17);*

(ii) *if* $4a = b^2$ *then*

$$s(w^n(a/b)) = \frac{b}{2n+2}, \quad n = 0,1,2,\dots \, . \tag{20}$$

In what follows we shall prove that the eigenvalue λ obtained in Theorem 7.4 is the unique point of the spectrum of A in a certain neighbourhood of a_{11}.

7.8 Proposition *Under the hypotheses of Theorem 7.4 we have:*

(i) *if* $4a < b^2$, *then* λ *is the unique point of the spectrum of* A *in the open disc* $U = \{z; |z - a_{11}| < \frac{1}{2} (b + (b^2 - 4a)^{1/2})\}$.

(ii) *if* $4a = b^2$, *then* λ *is the unique point of the spectrum of* A *in the closed disc* $V = \{z; |z - a_{11}| \leq \frac{1}{2} b\}$.

<u>Proof</u> Set $\lambda = a_{11} + w_1$ and let $\mu = a_{11} + w_2$ be another point of the spectrum of A. Suppose $|w_2| < b$. The open disc $\{z; |z - a_{11}| < b\}$ is contained in the resolvent set of C so that, according to Lemma 7.1, we have

$$w_i = \langle (w_i - B)^{-1} u, v \rangle, \quad i = 1,2. \tag{21}$$

Hence we may write

$$|w_1 - w_2| \leq |(w_1 - B)^{-1} - (w_2 - B)^{-1}| \; |u| \; |v| \leq \frac{a|w_1 - w_2|}{d(w_1 - B)d(w_2 - B)}. \tag{22}$$

From Theorem 7.4 we have $|w_1| \leq \frac{1}{2} (b - (b^2 - 4a)^{1/2}) =: p_1$. If we suppose $|w_2| < \frac{1}{2} (b + (b^2 - 4a)^{1/2}) =: p_2$ then

$$d(w_1 - B) \; d(w_2 - B) \geq (b - |w_1|)(b - |w_2|) > (b - p_1)(b - p_2) = a.$$

In this case it follows from (22) that $w_1 = w_2$. This completes the proof of point (i).

Let us suppose now $4a = b^2$. Let (z_n) be the sequence generated by (8') and let s_n be given by (18). We have

$$\lim_{n \to \infty} z_n = w_1; \quad |z_n| \leq s_n \quad n = 0,1,2,\ldots .$$

If $|w_2| \leq \frac{1}{2} b$ then we have successively

$$|w_2 - z_{n+1}| = |\langle ((w_2 - B)^{-1} - (z_n - B)^{-1}) u, v \rangle|$$

$$\leq \frac{|u| \; |v| \; |w_2 - z_n|}{d(w_2 - B)d(z_n - B)} \leq \frac{b^2 |w_2 - z_n|}{4(b - |w_2|)(b - |z_n|)}$$

$$\leq \frac{n + 1}{n + 2} |w_2 - z_n| \leq \ldots \leq \frac{1}{n + 2} |w_2|.$$

Hence $w_1 = w_2$. $\square$

Considering again the 2×2 matrix

$$A = \begin{pmatrix} 0 & a^{1/2} \\ -a^{1/2} & b \end{pmatrix}$$

we see that the uniqueness result obtained above is optimal within the class of problems depending on the parameters a and b. In Proposition 1 we have proved that the results obtained in Theorem 1 are also optimal within this class. It is to be expected, however, that the estimates could be improved by considering the same problem in a class defined by a greater number of parameters. We intend to exhibit now analogous estimates given this time in terms of three parameters; it is natural that they will be better in general than the ones given above since we are starting with a better information about the problem considered.

Theorem 2 *Let E be a Banach space written as a direct sum $E = E_1 \oplus E_2$ where E_1 is one-dimensional. Let $a_{11} \in C$, $u \in E_2$, $v \in {}^dE_2$ and $C \in B(E_2)$ be given. Suppose the linear operator $B = C - a_{11}$ is invertible and set*

$$a = |u| \, |v|, \quad b = d(B), \quad c = |1 + \langle B^{-2}u,v \rangle|.$$

If $3a \le b^2 c$, then:

(i) the iterative algorithm

$$\lambda_0 = a_{11}, \quad \lambda_{n+1} = a_{11} + \frac{\langle B^{-1}u,v \rangle}{-1 + \langle B^{-1}(\lambda_n - C)^{-1}u,v \rangle}, \quad n = 0,1,2,\ldots \quad (23)$$

is meaningful and it produces a convergent sequence $(\lambda_n)_{n \ge 0}$ whose limit λ is a simple eigenvalue of the operator $A \in B(E)$ given by (1).

(ii) The following estimates hold

$$|a_{11} - \lambda| \le \frac{1}{2}(a - m_0), \qquad (24)$$

$$|\lambda_{n+1} - \lambda_n| \le w^n(\tfrac{a}{bc}), \qquad (25)$$

$$|\lambda_n - \lambda| \le s(|\lambda_n - \lambda_{n-1}|) - |\lambda_n - \lambda_{n-1}| \le s(w^n(\tfrac{a}{bc})), \qquad (26)$$

where

$$p = \frac{ab}{a + b^2c} \; , \; q = \frac{b^3c}{m + b^2c}, \; m_0 = (b^2 - 4bp)^{1/2}, \tag{27}$$

$$m(r) = [(b + r)^2 - 4p(b + r)]^{1/2} = [(q-p+r)^2 - 4p^2]^{1/2}, \tag{28}$$

$$w(r) = r \, \frac{q - p + r - m(r)}{q - p - r + m(r)} \; , \tag{29}$$

$$s(r) = \frac{1}{2} (r + m(r) - m_0). \tag{30}$$

<u>Proof</u> With the substitution $z_n = \lambda_n - a_{11}$ the iterative algorithm (23) becomes

$$z_0 = 0, \; z_{n+1} = G(z_n), \; n = 0,1,2,\ldots \tag{23'}$$

where

$$G(z) = \frac{\langle B^{-1}u,v\rangle}{-1 + \langle B^{-1}(z - B)^{-1}u,v\rangle} \; . \tag{31}$$

Using the identity

$$(B - z)^{-1} = B^{-1} + z \, B^{-1} (B - z) \tag{32}$$

we deduce that the relation

$$z = G(z)$$

is equivalent to

$$z + \langle (B - z)^{-1}u,v\rangle = 0.$$

According to Lemma 1 it follows that if z^* is a fixed point of G then $\lambda = z^* + a_{11}$ is a simple eigenvalue of A. Thus our task is to prove that the iterative algorithm (23') is well defined and that the sequence produced by it is convergent. To this effect we shall use Proposition 1.4. We attach to the iterative algorithm (23') the function w given in (29) and the following approximate sets

$$Z(r) = \{z \in C; \; |z - G(z)| \leq r, \; |z| \leq b - h(r)\} \tag{33}$$

where

$$h(r) = s(r) + \frac{1}{2}(b + m_0) = \frac{1}{2}(b + r + m(r)). \tag{34}$$

First let us remark that the function w given by (29) is a rate of convergence on the interval $T = (0,\infty)$ and its estimate function is given by (30). This can easily be proved using Proposition 1.1.

We also remark that the condition $|z| \leq b - h(r)$ appearing in the definition of $Z(r)$ assures the fact that $G(z)$ is meaningful i.e. $B - z$ is invertible and

$$-1 + \langle B^{-1}(z-B)^{-1}u,v \rangle \neq 0. \tag{35}$$

The invertibility of $B - z$ is obvious because

$$d(B - z) \geq d(B) - |z| \geq b - (b - h(r)) = h(r) > 0.$$

Using the identity (32) we may write

$$|-1 + \langle B^{-1}(z-B)^{-1}u,v \rangle| = |1 + \langle B^{-2}u,v \rangle + z\langle B^{-1}(B-z)^{-1}B^{-1}u,v \rangle|$$

$$\geq \hat{c} - |z\langle B^{-1}(B-z)^{-1}B^{-1}u,v \rangle| \geq c - \frac{a|z|}{b^2(b-|z|)} \geq c - \frac{a(b-h(r))}{b^2 h(r)}.$$

It follows that (35) will be satisfied if

$$h(r) > \frac{ab}{a + b^2 c} = p.$$

The latter inequality can be derived immediately because the function h is increasing and $h(0) = \frac{1}{2}(b + m_0) > p$.

Now taking $r_0 = \frac{a}{bc}$ we have $|G(0)| \leq r_0$ and $h(r_0) = b$. This shows that $z_0 \in Z(r_0)$ for $z_0 = 0$ so that condition (i) of Proposition 1.4 is satisfied. Condition (ii) of that proposition will be also satisfied if we prove that $z \in Z(r)$ implies $z' = G(z) \in Z(w(r))$. The function h differs from the estimate function z only by an additive constant so that it satisfies the relation $h(w(r)) = h(r) - r$. Thus from $z \in Z(r)$ it follows

$$|z'| \leq |z| + |z' - z| = |z| + |G(z) - z|$$

$$\leq b - h(r) + r = b - h(w(r)).$$

Using the identity $A^{-1} - B^{-1} = A^{-1}(B - A)B^{-1}$ we can write

$$z' - G(z') = G(z) - G(z')$$

$$= \langle B^{-1}u,v\rangle \; \frac{\langle B^{-1}[(B - z)^{-1} - (B - z')^{-1}]u,v\rangle}{(1 + \langle B^{-1}(B - z)^{-1}u,v\rangle)(1 + \langle B^{-1}(B - z')^{-1}u,v\rangle)}$$

$$= \langle B^{-1}u,v\rangle \; \frac{(z' - z)\langle B^{-1}(B - z)^{-1}(B - z')^{-1}u,v\rangle}{(1 + \langle B^{-1}(B - z)^{-1}u,v\rangle)(1 + \langle B^{-1}(B - z')^{-1}u,v\rangle)} \; .$$

By virtue of the inequalities

$$|\langle B^{-1}u,v\rangle| \leq \frac{a}{b}, \quad |(B - z)^{-1}| \leq \frac{1}{d(B - z)} \leq \frac{1}{h(r)},$$

$$|1 + \langle B^{-1}(B - z)^{-1}u,v\rangle| \geq c - \frac{a(b - h(r))}{b^2 h(r)}$$

and of the corresponding inequalities for z' we obtain

$$|z' - G(z')| \leq \frac{r\,a^2\,b^2}{[(b^2 c + a)h(r) - ab][(b^2 c + a)(h(r)) - r) - ab]}$$

and it is easy to check that the right-hand side of the above inequality coincides with w(r).

At this stage conditions (i) and (ii) of Proposition 1.4 are satisfied so that the sequence $(z_n)_{n\geq 0}$ produced by (23') will converge to a point z_* and the following relations will be satisfied:

$$z_n \in Z(w^n(r_0)) \tag{36}$$

$$|z_n - z^*| \leq s(w^n(r_0)). \tag{37}$$

Using the fact that h is an increasing function from (36) we deduce that $z_{n-1} \in Z(|z_n - z_{n-1}|)$. The proof is complete. □

Let us consider again the numbers

$$a = |u|\,|v|, \quad b = d(B), \quad c = |1 + \langle B^{-2}u,v\rangle|$$

appearing in the statements of Theorems 1 and 2. We have clearly

$$c \geq 1 - a/b^2$$

so that

$$b^2 c - 3a \geq b^2 - 4a.$$

This shows that if the hypothesis of Theorem 1 is satisfied then the hypothesis of Theorem 2 is also satisfied. Moreover we have

$$\frac{1}{2}(a - m_0) \leq \frac{1}{2}(b - (b^2 - 4a))^{1/2}$$

so that the estimate (24) is sharper than the estimate (7).

The problem of finding the radius of an inclusion disc for the almost decomposable operators discussed above was considered first in a joint paper of M. Fiedler and V. Pták [14]. The rate of convergence of the corresponding iterative process was established in the Gatlinburg Lecture [49]. This was the first instance where the rate of convergence was computed establishing first the estimate function and using the functional equation. The inclusion disc expressed in terms of the more refined information given by the three parameters appears first in [13]. The treatment of this case based on the nondiscrete induction method is due to M. Kubr.

8 Factorization theorems

Under certain conditions it is possible to prove that an element a of a Banach algebra A may be written as a product of two elements of the algebra A, a = xy with x,y ∈ A. Results of that type are called factorization theorems and are important in harmonic analysis and in the theory of group representations.

First of all, it is obvious that a factorization a = xy may be written down immediately if the algebra A has a unit e; it suffices to take x = e and y = a. Hence we shall restrict our attention to Banach algebras without a unit. It is not difficult to give examples of Banach algebras where a factorization is not possible for every element; we shall discuss such an example later. Hence factorization theorems are not valid in general Banach algebras without supplementary conditions. It turns out that a suitable condition on the algebra which guarantees the possibility of factorization is the existence of a bounded approximate identity. Under this condition a factorization may be obtained for every element of the algebra using an iterative construction.

8.1 <u>Lemma</u> *Let A be a Banach algebra, let E be a subset of A and let b be a positive number such that $|u| \leq b$ for every u ∈ E. Then the following two conditions are equivalent:*

1. *for every x ∈ A and every $\varepsilon > 0$ there exists a u ∈ E such that*

$$|x - ux| < \varepsilon \tag{1}$$

2. *for every finite sequence $x_1, \ldots, x_n$ ∈ A and every $\varepsilon > 0$ there exists a u ∈ E such that*

$$|x_j - ux_j| < \varepsilon \tag{2}$$

for j = 1,2,...,n.

<u>Proof</u> Let us prove that condition 1. implies the formally stronger condition

2. To see that, it will be convenient to use the following abbreviation: the formal expression $(1-u)x$ will stand for the difference $x-ux$ so that left multiplication by $(1-u)$ is a meaningful operation which takes elements of A again into elements of A even though $1-u$ is not an element of A if A does not possess a unit. Now suppose condition 1. fulfilled and consider a finite sequence $x_1,\ldots,x_n \in A$ and an $\varepsilon > 0$. Let σ be a positive number to be determined later. According to our assumption, there exists a $u_1 \in E$ such that $|(1 - u_1)x_1| < \sigma$. For the same reason, there exists a $u_2 \in E$ with $|(1 - u_2)(1 - u_1)x_2| < \sigma$. Repeating this process, we obtain an $u_3 \in E$ with $|(1 - u_3)(1 - u_2)(1 - u_1)x_3| < \sigma$ and a sequence $u_1,\ldots,u_n \in E$ such that

$$|(1 - u_k)(1 - u_{k-1}) \cdots (1 - u_1)x_k| < \sigma \tag{3}$$

for $k = 1,2,\ldots,n$. Let us denote by v the element of A for which $1 - v = (1 - u_n)(1 - u_{n-1}) \cdots (1 - u_1)$. First of all let us show that

$$|(1 - v)x_k| < \sigma(1 + b)^{n-1} \tag{4}$$

for every $k = 1,2,\ldots,n$. Indeed,

$$|(1-v)x_k| \le |(1-u_n)\ldots(1-u_{k+1})|\ |(1-u_k)\ldots(1-u_1)x_k|$$

$$\le (1 + b)^{n-k}\sigma. \tag{5}$$

Now let $u \in E$ be such that $|(1 - u)v| < \sigma$. Since

$$(1 - u) = (1 - u)(1 - v) + (1 - u)v \tag{6}$$

we have, for each k

$$|(1 - u)x_k| \le |1 - u|\ |(1 - v)x_k| + |(1 - u)v|\ |x_k|$$

$$\le (1 + b)(1 + b)^{n-1}\sigma + \sigma m \tag{7}$$

where $m = \max(|x_1|,\ldots,|x_n|)$. Thus it suffices to choose σ small enough to have $((1 + b)^n + m)\sigma < \varepsilon$. The equivalence is established. $\quad\square$

8.2 <u>Definition</u> *Let A be a Banach algebra and E a subset of A bounded by b in norm. We shall say that E is a left approximate unit bounded by b if E satisfies one of the conditions of the preceding lemma.*

It turns out that considerably more powerful factorization theorems may be obtained by considering A-modules; the changes that have to be done in the proofs are essentially of a formal character. We shall see later that the stronger theorems thus obtained have very important consequences.

Let F be a Banach space which is a left A-module, $|ax| \le |a| \, |x|$ for all $a \in A$, $x \in F$. We shall denote by F_0 the closure in F of AF. If E is a bounded left approximate unit of A, it is easy to show that:

3. *for every pair of finite sequences* $a_1,\ldots,a_n \in A$ *and* $x_1,\ldots,x_m \in F_0$ *and every* $\varepsilon > 0$ *there exists an* $e \in E$ *such that*

$$|ea_i - a_i| < \varepsilon \text{ for } i = 1,2,\ldots,n$$

$$|ex_j - x_j| < \varepsilon \text{ for } j = 1,2,\ldots,m. \tag{8}$$

If A is a Banach algebra without a unit, we denote by B its unitization. The multiplicative linear functional on B which has A as its kernel will be denoted by f. The mapping P defined by $Pz = z - f(z)$ is a projection of B onto A. The set of all invertible elements of B will be denoted by G(B).

If F is a Banach left A-module with $|xy| \le |x| \, |y|$ for $x \in A$ and $y \in F$ then F is also a left B-module in an obvious manner; the above inequality remains valid for $x \in B$ as well. If A has a left approximate unit (not necessarily bounded) the following formula shows that $(By)^- = (Ay)^-$ for each $y \in F$.

If $x \in B$ then, for each $e \in A$,

$$xy = (Px + f(x)e)y - f(x)(ey - y). \tag{9}$$

The first summand lies in Ay and the second may be made arbitrarily small by a suitable choice of $e \in A$.

Now we are ready to formulate the main result. It consists of proving that, given an algebra A with a bounded left approximate unit, every element of AF^- may be factorized, in other words, in proving the identity $AF^- = AF$. More precisely, we have the following.

8.3 <u>Theorem</u> *Let* A *be a Banach algebra with a left approximate unit of norm*

b. *Let X be a Banach space which is a left A-module. Let y ∈ X be given.*
Then the following two conditions are equivalent:

1. *y lies in the closure of* AX

2. *for every ε > 0 there exists a z ∈ X and an a ∈ A such that*

$$y = az, \quad |a| \leq b, \quad |z - y| \leq \varepsilon; \tag{10}$$

moreover, the element z may be taken in the set $(Ay)^-$.

<u>Proof</u> For each positive $r \leq 1$ set

$$M(r) = \{[a,z] \in A \times (Ay)^-; \ a + r \in G(B), \tag{11}$$
$$(a + r)z = y, \ |z - y| \leq \varepsilon(1 - r), \ |a| \leq b(1 - r)\}.$$

Observe that

$$[0,y] \in M(1). \tag{12}$$

We intend to show that there exist two positive numbers c and d, $0 < c < 1$,
such that

$$M(r) \subset U(M(cr),dr) \tag{13}$$

for all positive $r \leq 1$.
    It will turn out later that

$$c = \frac{2b + 1}{2b + 2} \text{ and } d = (1 - c)b$$

is a possible choice.
    From Proposition 1.8 it will follow then that

$$M(t) \subset U(M(0), \frac{d}{1 - c} t) \tag{14}$$

for all positive $t \leq 1$.
    In particular, from (12) and (14), it follows that M(0) is nonvoid and it
is easy to see that if $[a,z] \in M(0)$ then (10) is satisfied.
    Let us prove now that (13) is satisfied for any r, $0 < r \leq 1$. Let e be
an arbitrary element of A such that $|e| \leq b$. Set $a' = a + (1 - c)re$, where
c is a number between 0 and 1 to be determined later. We have:

135

$$|a' - a| \leq (1 - c)br. \tag{15}$$

We have $a' + cr = a + r + (1 - c)r(e - 1) = (1+(1-c)r(e-1)(a+r)^{-1})(a + r) = (1 + u)(a + r)$ if we write u for

$$(1 - c)r(e - 1)(a + r)^{-1}.$$

To estimate u, write

$$u = (1-c)\ r(e-1)\ ((a+r)^{-1} - r^{-1}) + (1-c)(e-1).$$

If we choose c so as to have

$$(1 - c)(b + 1) \leq \frac{1}{2}$$

and then e for which

$$(1 - c)\ |(e - 1)((a + r)^{-1} - r^{-1})| \leq 4^{-1} \tag{16}$$

we shall have

$$|u| \leq \frac{3}{4} \text{ and } |(1 + u)^{-1}| \leq 4.$$

Since $a' + cr = (1 + u)(a + r)$ it follows that $a' + cr$ is invertible and

$$(a' + cr)^{-1} - (a + r)^{-1}$$

$$= (a' + cr)^{-1}(a' + cr - (a + r))(a + r)^{-1}$$

$$= -(a' + cr)^{-1}(1 - c)r(e - 1)(a + r)^{-1}$$

$$= -(a' + cr)^{-1}u = -(a + r)^{-1}(1 + u)^{-1}u.$$

Now

$$|(a' + cr)^{-1}y - y| \leq |(a + r)^{-1}y - y|$$

$$+ |((a' + cr)^{-1} - (a + r)^{-1}y| \leq \varepsilon(1 - r)$$

$$+ |(a + r)^{-1}(1 + u)^{-1}uy| \leq \varepsilon(1 - r)$$

$$+ 4|(a + r)^{-1}||uy|.$$

Finally choose e to have (16) and at the same time

$$4|(a + r)^{-1}| \ |(e - 1) (a + r)^{-1}y| < \varepsilon. \tag{17}$$

It will follow that

$$|(a' + cr)^{-1}y - y| \leq \varepsilon(1 - r) + \varepsilon(1 - c)r = \varepsilon(1 - cr).$$

Taking $z' = (a' + cr)^{-1}y$ we have thus

$$[a',z'] \in M(cr).$$

Also, $|z' - z| \leq (1 - c)\varepsilon r$ so that, for $\varepsilon \leq b$

$$[a,z] \in U(M(cr),dr).$$

Hence (13) holds with $d = (1 - c)b$. The proof is complete. □

To illustrate the power of the factorization theorem let us state the following proposition which has important applications and which is an immediate corollary of the preceding theorem.

8.4 Proposition *Let A be a Banach algebra with a bounded left approximate unit. If $x(n) \in A$ is a sequence such that $\lim x(n) = 0$ then there exists an $a \in A$ and a sequence $y(n) \in A$ such that $x(n) = a \ y(n)$ for all n and $\lim y(n) = 0$.*

Proof Denote by X the Banach space of all sequences $x:N \to A$ such that $\lim x(n) = 0$, the norm being defined as $|x| = \max_n |x(n)|$. Then X is a left A-module if the action of A on X is defined by $(ax) (n) = a \ x(n)$, $n \in N$. Now it suffices to apply the factorization theorem to the pair A and X. □

The possibility of simultaneous factorizations of sequences tending to zero plays an important role in proving that the algebraic properties of certain mappings of Banach algebras already imply their continuity.

Let us first consider the continuity of positive functionals in Banach star-algebras. A Banach algebra with involution, or a Banach star-algebra for short, is a Banach algebra such that, for every $x \in A$ an element x^* is given such that the mapping $x \to x^*$ satisfies the following conditions

$$(x + y)^* = x^* + y^*$$

$$(\alpha x)^* = \alpha^* x^*$$

$$(xy)^* = y^* x^*$$
$$x^{**} = x$$

for all $x, y \in A$ and all $\alpha \in \mathbb{C}$.

Such a mapping is called an involution. Observe that no condition conn-ecting the involution with the topology of A is imposed, in particular con-tinuity of the involution is not assumed.

It is interesting to observe, however, that continuity of the involution may be proved if another algebraic condition is added, namely if A is semi-simple. Although this result is not our main concern here, we mention it at this point since it represents another instance of a situation where purely algebraic assumptions already imply its continuity. In this section we con-centrate on automatic continuity results based on the possibility of factor-ization.

In this section we intend to discuss continuity of positive functionals and of multipliers. In both cases the preceding corollary of the factori-zation theorem represents the crucial point of the proof.

A linear functional f on a Banach star-algebra A is said to be positive if $f(x^*x) \geq 0$ for every $x \in A$.

Without supplementary conditions positive functionals need not be con-tinuous. We intend to show, however, that every positive functional is con-tinuous on A if A has a bounded two-sided approximate identity.

8.5 Definition *A *-representation of a Banach star-algebra* A *on the Hilbert space* H *is an algebraic homomorphism*

$$\pi: A \to B(H)$$

which preserves the involution, in other words

$$\pi(x^*) = \pi(x)^*$$

for every $x \in A$.

We shall use the following automatic continuity result.

8.6 Theorem *Let* A *be a Banach star-algebra and let* π *be a star represent-ation of* A *in some Hilbert space* H. *Then* π *is continuous.*

The second step consists in the construction for a given positive

138

functional f, of a certain representation.

Given a positive functional f on a Banach star-algebra A, it follows from the Cauchy-Schwarz-Bunyakovskii inequality that

$$\{x \in A; f(Ax) = 0\} = \{x \in A; f(x^*x) = 0\}.$$

This set is clearly a left ideal and will be denoted by L_f. Denote by Q_f the canonical quotient map of A onto $A - L_f$. Define, on $A - L_f$, a scalar product by the formula

$$(Q_f(a), Q_f(b)) = f(b^*a)$$

and denote by H_f the completion of the praehilbert space $A - L_f$. Then a representation π_f of A on H_f may be constructed as follows. Given a $\in$ A consider the mapping of the praehilbert space $A - L_f$ into itself which assigns to every $Q_f(x)$ the class $Q_f(ax)$. Clearly this mapping is bounded. We denote by π_f its extension by continuity to the whole of H_f. Then π_f is a *-representation of A in H_f.

Now we are able to prove the following remarkable fact.

8.7 **Lemma** *Let A be a Banach star-algebra and let f be a positive functional on A.*

If b $\in$ A is given, denote by f_b the linear functional defined by

$$f_b(x) = f(b^*x\ b).$$

Then f_b is again a positive functional and f_b is continuous.

Proof The proof of the first assertion is straightforward. To prove the second assertion, consider the representation π_f. We have

$$f_b(x) = f(b^*xb) = (Q_f(xb), Q_f(b)) = (\pi_f(x)\ Q_f(b), Q_f(b))$$

so that

$$|f_b(x)| \le |\pi_f(x)|\ |Q_f(b)|^2 \le |\pi_f(x)|\ f(b^*b).$$

Since π_f is continuous we have thus proved the continuity of f_b. □

Now everything is ready for the proof of the main result.

8.8 Theorem *Let A be a Banach star-algebra with a bounded two-sided approx-imate unit. Then every positive functional on A is continuous.*

<u>Proof</u> Consider a positive functional f on A. The proof consists of two steps. First we prove the continuity of functionals of the form $x \to f(bxc)$ for fixed $b, c \in A$. This is an immediate consequence of the following polarization identity

$$4f(b \cdot c) = f_{c+b*} - f_{c-b*} + i\, f_{c+ib*} - i\, f_{c-ib*}$$

and of the preceding proposition.

Now suppose that $x_n \in A$ and $\lim x_n = 0$. Since A has a bounded left approximate unit, we may write $x_n = by_n$ for some $b \in A$ and a sequence y_n with $\lim y_n = 0$. Since A has a bounded right approximate unit there exists a $c \in A$ and a sequence z_n with $y_n = z_n c$ and $\lim z_n = 0$. Thus $x_n = bz_n c$ and $\lim f(x_n) = \lim f(b\, z_n\, c) = 0$ since $f(b \cdot c)$ is continuous. □

Continuity of multiplers is another instance of a result based on the possibility of factorizing sequences tending to zero.

Let A be a Banach algebra. For each $a \in A$ let us denote by $L(a)$ the mapping $L(a): A \to A$ defined by

$$L(a)x = ax, \quad x \in A.$$

Clearly $L(a)$ is a bounded linear operator for each $a \in A$ and its norm satisfies $|L(a)| \leq |a|$.

Consider now a linear mapping T of A into itself such that T commutes with all operators $L(a)$, in other words

$$T(ax) = a\, T(x)$$

for every pair, $a, x \in A$. We intend to show that T is automatically continuous if A possesses a bounded right approximate unit. Given a sequence x_n tending to zero we may write x_n in the form $x_n = y_n b$ for some $b \in A$ and some y_n with $\lim y_n = 0$. It follows that

$$T(x_n) = T(y_n a) = y_n T(a) \to 0$$

and this completes the proof.

We have seen, in the proof of the factorization theorem, that boundedness of the approximate unit was used in an essential manner. Let us show now that boundedness of the approximate unit is essential for the result itself not just for the proof.

8.9 <u>Example</u> Let A be the algebra of all bounded complex valued functions f defined on the set $T = \{t; 1 \le t\}$ and such that $\lim t\, f(t) = 0$ as t tends to infinity, the operations being defined pointwise. The algebra A becomes a Banach algebra under the norm

$$|f| = \sup |t\, f(t)|.$$

If $f \in A$ may be factorized, $f = f_1 f_2$ say, then

$$t^2\, f(t) = t\, f_1(t)\, tf_2(t)$$

tends to zero so that factorizable elements represent a proper subset of A. It follows that A cannot possess a bounded approximate unit. There is, however, an obvious unbounded one: consider, for each $m > 1$, the characteristic function $e(m)$ of the interval $1 \le t \le m$. The set

$$E = \{e(m) : m \ge 1\}$$

is an approximate unit for A in the following sense: for each $x \in A$ and each $m > 1$ we have

$$|e(m)x - x| = \sup_{t>m} |t\, x(t)|$$

so that $|e(m)x - x|$ may be made arbitrarily small if m is large enough - at a price though, the norm of $e(m)$ being exactly m. If y is the function

$$y(t) = t^{-2}$$

then $|e(m)y - y| = \frac{1}{m}$ so that, to make $|e(m)y - y|$ less than ε we have to take an $e(m)$ of norm at least $1/\varepsilon$. ▫

Recently the factorization theorem has been extended to a somewhat stronger statement concerning power factorizations [1]. Let us conclude this section by remarking that a strengthening of the power factorization theorem as well as a simplification of the proof may be obtained by adapting the nondiscrete induction argument given above to this more complicated situation. The result [55] is as follows:

Theorem *Let A be a Banach algebra with a left approximate unit of norm* β. *Let F be a Banach space which is a left A-module. Let* a_n *be an arbitrary sequence of positive numbers. Let* $y(1), y(2),\ldots$ *be a sequence of elements of* AF^- *such that*

$$\lim a_n^{-1} |y(n)|^{1/n} = 0$$

and let $\epsilon > 0$ *be given.*

Then there exists a sequence $z(n) \in Ay(n)^-$ *and an element* $a \in A$ *such that*

$$y(n) = a^n z(n) \text{ for all } n \in \mathbb{N},$$

$$|a| \leq \beta.$$

$$|z(n) - y(n)| \leq a_n^n \epsilon^n \text{ for all } n \in \mathbb{N}.$$

9 Transitivity in operator algebras

There is an important result in the theory of C^*-algebras the proof of which requires an inductive construction: for representations of C^*-algebras *topological irreducibility and strict irreducibility are equivalent*. This result is a corollary of a much stronger transitivity theorem - Section 8.2 of the monograph of J. Dixmier. In this section we shall present a neat proof as well as an improvement of this classical theorem.

Let H be a Hilbert space, B(H) the algebra of all bounded linear operators on H. Let A be a self-adjoint norm-closed subalgebra of B(H). We shall make the following two assumptions about A.

1. $AH \neq 0$

2. the only closed subspaces of H invariant with respect to A are 0 and H;
    (such algebras are called transitive subalgebras of B(H)).

Let us prove first that Ax is dense in H for every $x \neq 0$. To see this, set

$$M = \{x \in H; \; Ax = 0\}.$$

Clearly M is a closed subspace of H invariant with respect to A. It follows from assumption 1. that M cannot be the whole of H so that M = 0. Thus Ax contains a nonzero element whenever $x \neq 0$. The closure of Ax is thus a closed A-invariant subspace different from the zero subspace so that $(Ax)^- = H$. Let us prove further that the commutant A' of consists of scalar multiples of the identity only. To see this, consider an orthogonal projection $P \in A'$ and let us show that P is either zero or the identity. Indeed, if P is different from the zero projection, there exists an $x \neq 0$, x = Px. Hence

$$Ax = APx = PAx \text{ for every } A \in A$$

so that the range of P contains the whole subspace Ax. Since $x \neq 0$ the subspace Ax is dense in H so that P = I.

Since every W^*-algebra is generated by its projections, it follows that A' consists of multiples of the identity only; moreover, it follows from the double commutant theorem that A is dense in B(H) in the strong operator topology.

In other words: given an arbitrary $T \in B(H)$, a finite set of vectors $x_1,\ldots,x_n \in H$ and a positive number ε then there exists an $A \in A$ such that $|Ax_i - Tx_i| < \varepsilon$ for $i = 1,2,\ldots,n$. According to the Kaplansky density theorem an $A \in A$ may be chosen whose norm does not exceed the norm of T.

We intend to show that there exists an operator $B \in A$ such that

$$Bx_i = Tx_i \text{ for } i = 1,2,\ldots,n$$

and whose norm does not exceed $|T| + \varepsilon$.

This result improves a classical fact (as a reference, let us give Chapter 2.8 of the monograph of J. Dixmier).

Given n linearly independent vectors $u_1,\ldots,u_n$ and n vectors $v_1,\ldots,v_n$ we shall denote by $W(u_1,\ldots,u_n; v_1,\ldots,v_n)$ the linear operator $W \in B(H)$ defined by the requirements

$$Wu_j = v_j \text{ for } j = 1,2,\ldots,n$$

$$Wx = 0 \quad \text{for } x \perp u_1,\ldots,u_n.$$

We shall show that W has minimal norm among all operators $T \in B(H)$ for which $Tu_j = v_j$. To see this, denote by P the orthogonal projection onto the subspace generated by $u_1,\ldots,u_n$. Observe that $W = WP$. Suppose T satisfies $Tu_j = v_j$ for $j = 1,2,\ldots,n$. Then

$$|T| \geq \sup \{|Tx \; ; \; x = Px, \; |x| \leq 1\}$$

$$= \sup \{|Wx \; ; \; x = Px, \; |x| \leq 1\}$$

$$= \sup \{|WPy \; ; \; |y| \leq 1\} = |WP| = |W|.$$

Now we can state the main result.

9.1 <u>Theorem</u> *Let A be a strongly dense* C^*-*algebra of* $B(H)$. *Suppose we are given* n *linearly independent vectors* $u_1,\ldots,u_n \in H$ *and* n *vectors* $v_1,\ldots,v_n \in H$. *Then, for each* $\varepsilon > 0$, *there exists an operator* $R \in A$ *such that* $Ru_j = v_j$ *for* $j = 1,2,\ldots,n$ *and* $|R| \leq |W(u_1,\ldots,u_n; v_1,\ldots,v_n)| + \varepsilon$.

<u>Proof</u> First of all, there exists a nonsingular n by n matrix M such that the

vectors

$$x_i = \sum_k m_{ik} u_k$$

form an orthonormal set: if we denote by $y_i = \sum m_{ik} v_k$ it follows from the nonsingularity of M that the requirement on a linear operator to satisfy $Tu_j = v_j$ is equivalent to the requirement that

$$Tx_j = y_j.$$

We introduce the following abbreviations.

$$|W(u_1,\ldots,u_n;\, v_1,\ldots,v_n)| = |W(x_1,\ldots,x_n;\, y_1,\ldots,y_n)| = \beta.$$

Choose a number $0 < \alpha < \min(1, \beta + \varepsilon)$ and set, for each $0 < r < 1$

$$M(r) = |A \in A;\ |A| \leq (\beta+\varepsilon)\,(1-r),$$

$$|Ax_j - y_j| \leq \alpha r \text{ for } j = 1,2,\ldots,n\}.$$

Suppose $A \in M(r)$; define an operator S by the formula

$$Sx = \sum_{j=1}^{n} (x,x_j)\,(y_j - Ax_j)$$

so that $S \in B(H)$ and $|S| \leq \alpha r$.

By the Kaplansky density theorem there exists a $T \in A$ such that

$$|T| \leq |S| \quad \text{and} \quad |Tx_j - Sx_j| < \alpha \omega r$$

for $j = 1,2,\ldots,n$ where

$$\omega = \frac{\beta + \varepsilon - \alpha}{\beta + \varepsilon}.$$

Set $B = A + T$. We have thus $B \in A$ and $|B - A| = |T| \leq \alpha r < r$. Further

$$|B| \leq |A| + |T| \leq (\beta+\varepsilon)\,(1-r) + \alpha r = (\beta+\varepsilon)(1-\omega r)$$

$$Bx_j - y_j = Ax_j - y_j + Tx_j = Tx_j - Sx_j$$

so that

$$|Bx_j - y_j| \leq \alpha \omega r.$$

It follows that $B \in M(\omega r)$ and $A \in U(M(\omega r),r)$. Now let r_0 be determined from the equation $\beta = (\beta+\varepsilon)(1-r_0)$. By the Kaplansky density theorem there exists an $A_0 \in A$ such that

$$|A_0| \le \beta \quad \text{and} \quad |A_0 x_j - y_j| \le \alpha r_0.$$

It follows that $A_0 \in M(r_0)$. Thus $\lim M(r)$ is nonvoid and any $R \in \lim M(r)$ satisfies the requirements of the theorem. □

To conclude, let us state the theorem on irreducible representations of C^*-algebras. We shall see that it uses only a fraction of the power of the preceding theorem.

9.2 Theorem *Every topologically irreducible representations of a C^*-algebra is already strictly irreducible.*

Proof If π is a representation of a C^*-algebra R in the Hilbert space H then it is known that $\pi(R)$ is again a C^*-algebra. Now suppose that π is topologically irreducible, in other words, that 0 and H are the only closed subspaces of H invariant with respect to $\pi(R)$. We have seen that this is equivalent to saying that, for each vector $x \ne 0$ the subspace $\pi(R)x$ is dense in H. It follows from the preceding theorem however, that $\pi(R)x$ is in fact the whole of H. We have thus strict irreducibility. □

10 Stability of openness and exactness

A typical example of an application of the Subtraction Theorem 2.2 is the stability of openness for linear mappings - a small perturbation of an open bounded linear mapping is again open. The induction theorem may be used to obtain an analogous proposition for exactness: a small perturbation of an exact sequence is again exact. A result of this type is needed to establish compactness of joint spectra in the theory of J.L. Taylor.

The constructions work in a much more general situation which we now proceed to explain.

Given a nonvoid set S in a Banach space E we denote by $P(S)$ the perturbation class associated with S; it is defined as the set of those elements $x \in E$ which may be added to an arbitrary element of S without leaving the set S:

$$P(S) = \{x \in E; \ x + S \subset S\}.$$

Clearly $P(S)$ is closed with respect to addition. The set $P(S)$ may be obviously represented as the intersection of all sets of the form $-s + S$, the element s ranging over the whole of S:

$$P(S) = \cap \{-s + S; \ s \in S\};$$

it follows from this representation that $P(S)$ is closed if S is closed. It is interesting but less obvious to observe that $P(S)$ is closed if S is open.

10.1 <u>Proposition</u> *If S is a nonvoid open set in a Banach space E then* $P(S)$ *is closed in* E.

<u>Proof</u> Suppose $x \in P(S)^-$ and let $s \in S$ be given. The set S being open, there exists an $\varepsilon > 0$ such that $U(s,\varepsilon) \subset S$. Since $x \in P(S)^-$ there exists a $y \in P(S)$ with $d(x,y) < \varepsilon$. It follows that $s' = s + (x - y) \in S$ so that $x + s = y + s + (x - y) = y + s' \in S$.

This completes the proof.   □

It is easy to see that $P(S)$ will be a linear space if S satisfies the following additional assumption: for each $\lambda \neq 0$ the set λS is contained in S.

The following shows - roughly speaking - that, under the condition that elements of S may be well approximated by elements of $P(S)$ then S itself is a subset of $P(S)$, possibly even of a proper subset of $P(S)$.

Closed sets

10.2 Proposition *Let F be a Banach space, S a subset of F. Let $B \subset F$ be closed with respect to addition and closed in F. Suppose that*

(i) $S - B \subset S$;

(ii) *there exists a β, $0 < \beta < 1$, such that for each $s \in S$, dist $(s,B) \leq \beta |s|$.*

Then $S \subset B$.

Proof Let $s \in S$ be fixed. For each $t > 0$ set

$$Z(t) = \{b \in B;\ |s - b| < \frac{1}{\beta + 1} t\}.$$

Given $b \in Z(t)$, we have $s - b \in S$ by (i), so that, by (ii), there exists a $b_1 \in B$ with $|s - b - b_1| \leq \beta |s - b| < \frac{\beta}{(\beta + 1)} t$. Set $b' = b + b_1$ so that $b' \in Z(\beta t)$ and

$$|b' - b| = |b_1| \leq |s - b - b_1| + |s - b| < t,$$

whence $b' \in Z(\beta t) \cap U(b,t)$. Since b was an arbitrary point of $Z(t)$, we have

$$Z(t) \subset U(Z(\beta t),t)$$

and the induction theorem applies. □

10.3 Corollary *Let F be a Banach space, S a subset of F. Let u be a continuous and open linear mapping of a Banach space E into F. Suppose that*

(i) $S + u(E) \subset S$,

(ii) *there exists a β, $0 < \beta < 1$, such that for each $s \in S$, dist $(s,u(E)) < \beta |s|$.*

Then $S = u(E)$.

<u>Proof</u> Since u is open, its range u(E) is closed in F, The set B = u(E)
satisfies the hypotheses of the preceding proposition. We have thus $S \subset u(E)$.
Take an arbitrary $t \in S$. Since $S \subset u(E)$ we have $t \in u(E)$ so that $- t \in u(E)$
as well. Hence

$$0 = t + (-t) \in S + u(E) \subset S$$

and $0 \in S$ implies $u(E) \subset S$. □

It is not difficult to see (since u(E) will be closed), that, in the
corollary, completeness of F is not essential. Indeed, we have

10.4 <u>Proposition</u> *Let F be a normed linear space, S a subset of F. Suppose
that u is a bounded linear and open mapping of a Banach space E into F such
that*

(i) $S + u(E) \subset S$,

(ii) *there exists a β, $0 < \beta < 1$, such that for each $s \in S$, dist $(s, u(E)) < \beta |s|$.
Then $S \subset u(E)$.*

<u>Proof</u> Consider a fixed $s \in S$. For each $t > 0$ set

$$W(t) = \{x \in E: |s - u(x)| \leq t\}.$$

Suppose that $x \in W(t)$. Since $s - u(x) \in S$ by (i), there exists an $x_1 \in E$
such that $|s - u(x) - u(x_1)| < \beta |s - u(x)| \leq \beta t$. Since u is open, we have
dist $(z, \text{Ker } u) < d|u(z)|$ for some $d > 0$ and all $z \in E$. Hence there exists
an $x_2 \in E$ such that $u(x_2) = u(x_1)$ and

$$|x_2| \leq d|u(x_1)| \leq d(|s-u(x) - u(x_1)| + |s-u(x)|) < d(\beta+1)t.$$

If $x' = x + x_2$, we have $|s-u(x')| < \beta t$ and $|x'-x| < d(\beta+1)t$. Since x was an
arbitrary element of W(t), we have proved the inclusion

$$W(t) \subset U(W(\beta t), d(\beta+1)t).$$

To conclude the proof, apply Proposition 1.8. □

10.5 <u>Remark</u> Another choice of the parametrization leads to a simpler
formula. Indeed, the sets

$$Z(s) = W\left(\frac{1}{d(\beta+1)} s\right)$$

satisfy

$$Z(s) \subset U(Z(\beta s),s)$$

for all positive s.

Stability of openness

Let us turn now to the study of open mappings. Before stating the results it will be convenient to introduce a measure of openness for linear mappings.

10.6 __Definition__ *Given a linear mapping* T *from a normed linear space* E *into a normed linear space* F, *set*

$$m(T) = \sup \{\text{dist } (x, \text{Ker } T); |Tx| \leq 1.\}.$$

This measure of openness has the following properties:

(i) *The number* m(T) *is finite if and only if* T *is open; in this case* m(T) *is the minimum of all numbers* α *with the following property*

$$\text{dist } (x, \text{Ker } T) \leq \alpha|Tx| \text{ for all } x$$

(ii) *If* m(T) *is finite then* $\frac{1}{m(T)}$ *is the supremum of those* $\beta > 0$ *for which*

$$\beta V \cap T(E) \subset TU;$$

here U *and* V *are the closed unit cells of* E *and* F, *respectively.*

(ii) $\frac{1}{m(T)} = d(\hat{T})$ *where* $\hat{T}$ *is the mapping from* E/Ker T *into* F *defined by* $\hat{T}(x + \text{Ker } T) = Tx.$

The proof of these statements is quite straightforward; as an example, let us check the second assertion.

Suppose that m(T) is finite and let us shown that for every $0 < \beta < \frac{1}{m(T)}$, the inclusion

$$\beta V \cap T(E) \subset TU$$

holds. Indeed, suppose $x \in E$ is such that $Tx \in \beta V$. Then $|T(\beta^{-1}x)| \leq 1$ so

150

that dist $(\beta^{-1}x, \text{Ker } T) \leq m(T)$, there exists, accordingly, a $k \in \text{Ker } T$ such that $|\beta^{-1}x - k| < \beta^{-1}$ whence

$$Tx = T(x - \beta k) \text{ and } x - \beta k \in U.$$

We see thus that the supremum of the numbers β is $\geq \frac{1}{m(T)}$.

On the other hand, let us prove that $\beta \leq \frac{1}{m(T)}$ for every positive β for which

$$\beta V \cap T(E) \subset TU.$$

Indeed suppose $x \in E$ satisfies $|Tx| \leq 1$. Then $T(\beta x) \in TU$ so that there exists a $y \in U$ with $T(\beta x) = T(y)$ whence $\beta x - y \in \text{Ker } T$. Thus

$$\text{dist } (x, \text{Ker } T) \leq |x - \frac{1}{\beta}(\beta x - y)| = \frac{1}{\beta}|y| = \frac{1}{\beta}$$

so that, by the definition of $m(T)$, $\frac{1}{\beta} \leq m(T)$.

Let us state now our result concerning the stability of openness:

10.7 __Theorem__ *Let E, F be two Banach spaces, u and u' two elements of $B(E,F)$. Suppose that u is open and that $m(u)|u'-u| < 1$. Then the following two implications hold:*

(i) *if Ker $u' \supset$ Ker u, then Ker $u' = $ Ker u;*

(ii) *if Range $u' \subset$ Range u, then u' is open and Range $u' = $ Range u.*

More precisely, if Range $u' \subset$ Range u, we have

$$m(u') \leq m(u)(1 - m(u)|u'-u|)^{-1}.$$

__Proof__ The first assertion is obvious. Indeed, suppose that Ker $u' \supset$ Ker u and that there exists an $x \in$ Ker $u' \smallsetminus$ Ker u. Since Ker u is closed, we have dist $(x, \text{Ker } u) > 0$. If z is an arbitrary element of Ker u, we have $z \in$ Ker u' as well; hence $u'(x - z) = 0$; at the same time

$$\text{dist } (x, \text{Ker } u) \leq m(u)|u(x)| = m(u)|u(x-z) - u'(x-z)|$$

$$\leq m(u)|u'-u||x-z|.$$

Since z was arbitrary in Ker u, it follows that

$$\text{dist } (x, \text{ Ker } u) \leqq m(u) \; |u' - u| \; \text{dist } (x, \text{ Ker } u)$$

which is impossible.

To prove the second assertion assume that Range $u' \subset$ Range u and take a $v \in u(E)$ with $|v| < 1$. It will be sufficient to assume $|u'-u| > 0$. We intend to show that $v = u'(x_0)$ for some $x_0 \in E$ of norm not exceeding

$$m(u) \; (1 - m(u) \; |u'-u|)^{-1}.$$

Set $\beta = m(u) \; |u'-u|$. For each $t > 0$, let

$$W(t) = \{x \in E; \; |u'(x) - v| < t\}.$$

Suppose $x \in W(t)$. Since $u'(E) \subset u(E)$, we have $u'(x) - v = u(z)$ for some z with $|z| < m(u)t$. Set $x' = x - z$. Then $|x' - x| < m(u)t$ and $|u'(x') - v| = |u'(x) - u'(z) - v| = |u(z) - u'(z)| \leq |u-u'| \; |z| < \beta t$. We have thus $x' \in W(\beta t)$. Since x was an arbitrary element of $W(t)$ we have proved the inclusion

$$W(t) \subset U(W(\beta t), m(u)t).$$

Setting $Z(s) = W(s/m(u))$, we have

$$Z(s) \subset U(Z(\beta s), s).$$

It follows from the induction theorem that

$$0 \in W(1) = Z(m(u)) \subset U \left(Z(0), \frac{m(u)}{1-\beta} \right).$$

so that there exists an $x_0 \in Z(0)$ with $|x_0| \leq m(u)/(1-\beta)$. Since $x_0 \in Z(0)$, we have $u'(x_0) = v$.

If we denote by $\overset{\circ}{V}$ the open unit ball of F and by U the closed unit ball of E, we have proved the inclusion

$$u(E) \cap \overset{\circ}{V} \subset u' \left(\frac{m(u)}{1-\beta} U \right). \qquad \qquad \Box$$

10.8 <u>Remark</u> If the inclusion: Range $u' \subset$ Range u is not satisfied, the condition $m(u) \; |u'-u| < 1$ alone does not guarantee the openness of u' nor the equality of the ranges.

10.9 Remark It is interesting to deduce the last result directly from the Subtraction Theorem. Denote by U and V the closed unit cells of E and u(E), respectively. Since u is open, we have $\frac{1}{m(u)} V \subset u(U)$. Since $u'(E) \subset u(E)$, it follows that

$$\frac{1}{m(u)} V \subset u'(U) + |u' - u|V.$$

If $|u'-u| < 1/m(u)$, we have, by Theorem 2.2, the inclusion

$$u'(U) \supset \left(\frac{1}{m(u)} - |u' - u| \right) V$$

so that u' is open as well and

$$\frac{1}{m(u')} \geq \frac{1}{m(u)} - |u'-u|.$$

Stability of exactness

10.10 Proposition *Let E, F, G be three Banach spaces, u, u' $\in$ B(E,F), v, v' $\in$ B(F,G). Suppose that the mappings u and v are open and that*

$$\text{Range } u = \text{Ker } v, \quad \text{Range } u' \subset \text{Ker } v'.$$

Then there exists an $\varepsilon > 0$ such that

$$|u'-u| < \varepsilon \text{ and } |v'-v| < \varepsilon$$

together imply

$$\text{Range } u' = \text{Ker } v'.$$

More precisely, it suffices to take ε for which

(iii) $\varepsilon(m(v) + m(u) (1 + m(v)\varepsilon)) < 1.$

Proof Let ε be a positive number which satisfies inequality (iii). There exist two numbers $h(u) > m(u)$ and $h(v) > m(v)$ such that

$$\beta = \varepsilon(h(v) + h(u) (1 + h(v)\varepsilon)) < 1.$$

Set S = Ker v' so that S + u'(E) $\subset$ S and assume that $|u' - u| < \varepsilon$ and

$|v-v'| < \varepsilon$. We shall show that

(iv) there exists an a such that for each $s \in S$ there exists an $x \in E$ with

$$|s - u'(x)| \le \beta|s|, \quad |x| \le a|s|.$$

Consider a fixed $s \in S$. Since v is open, there exists an $y \in F$ such that $v(y) = v(s)$ and $|y| \le h(v) \; |v(s)| = h(v) \; |(v-v')(s)| \le h(v)\varepsilon \; |s|$. Since $v(s-y) = 0$, there exists an $x \in E$ with $u(x) = s-y$ and

$$|x| \le h(u) \; |s-y| \le h(u)(|s| + |y|) \le h(u)(1 + h(v)\varepsilon) \; |s|.$$

Set $a = h(u) (1 + h(v)\varepsilon)$. Now

$$|s-u'(x)| \le |s-u(x)| + |(u'-u)x| \le |y| + \varepsilon|x| \le h(v)\varepsilon \; |s| + \varepsilon a|s| = \beta|s|.$$

The rest follows by considering the sets

$$W(t) = \{x \in E; \; |s-u'(x)| < t\}$$

which satisfy

$$W(t) \subset U(W(\beta t), at). \qquad \square$$

11 Interpolation spaces, approximate solutions and a theorem of J Moser

In the present section we intend to exhibit an application of the method of nondiscrete mathematical induction to a highly ingenious modification of the Newton process for an iterative construction of solutions of nonlinear partial differential equations. Surprisingly enough it turns out that - in the situation to be described below - it is more advantageous not to use the exact solution of the linearized equations at each step of the process but take instead a so-called approximate solution. A precise definition will be given later; for the moment it will be sufficient to say that the main requirement on the solution of the linearized equation

$$f(x_n) + f'(x_n) (x_{n+1} - x_n) = 0$$

is its smoothness so that it is preferable to take a sufficiently smooth x_{n+1} even at the expense of not satisfying the linearized equation exactly. The reason for this may be explained easily as follows: in many applications the sequence of approximations x_n is considered in a scale of spaces $E(r)$, r being some measure of smoothness (number of derivatives) and it may happen that the norm $|x_{n+1} - x_n|_r$ can only be estimated in terms of $|x_n - x_{n-1}|_{r+s}$, where s is a positive integer. If we were to use these estimates all derivatives would be exhausted after a finite number of steps. This leads to the idea of applying a smoothing operator at each step.

We intend to use our method to discuss a modification of the Newton method which uses the technique of smoothing operators and which was proposed by J. Moser.

We shall try to present the abstract core of the result eliminating thereby inessential technicalities: nevertheless, it will be necessary to explain the motivation of the definitions and results by means of some concrete examples. We shall work with a scale of Banach spaces with properties similar to those of Sobolev spaces. The basic Banach space E will be contained between two normed spaces

$$E_1 \subset E \subset E_0,$$

the injections being continuous.

We may think of E_0 as being an L_2-space; E_1 would be a subspace of smooth functions and E could be viewed as a sort of interpolation space intermediate between E_1 and E_0.

Before giving the definition of an approximate solution of a linear equation we shall consider the following particular case. A given element $x \in E$ may be approximated in the norm of E_0 arbitrarily well by smooth functions $y \in E_1$; it is to be expected, however, that the function y will have sharp bends so that its E_1 norm will be fairly large. It turns out that the relation between the degree of approximation and the norm of the approximating element may be used to characterize elements of E.

Before giving the exact statement of the theorem we intend to discuss two concrete examples of constructions of approximate solutions in the linear case. In order to do so it will be useful to describe in more detail a concrete scale of spaces which will serve to illustrate the notions to be used later. The discussion presented below follows lossely the exposition given in the lectures of J. Moser [24] with some more emphasis on the abstract substance of the notion of an approximate solution.

Let n be a fixed natural number. If k is a multi-index $k = (k_1,\ldots,k_n)$, the k_i being integers and if x is an n-tuple of real numbers $x = (x_1,\ldots,x_n)$ we write

$$(k,x) = \Sigma\, k_j x_j$$

$$|k|^2 = \Sigma\, k_j^2.$$

For a finite trigonometrical polynomial of the form

$$v = \underset{k}{\Sigma}\, v_k \exp i\,(k,x)$$

and a real number r define the norm $|v|_r$ by the formula

$$|v|_r^2 = \underset{k}{\Sigma}\, |k|^{2r}\, |v_k|^2.$$

The completion of the linear space of all trigonometrical polynomials with respect to the norm $|\cdot|_r$ will be denoted by $V(r)$. Clearly

$$V(r) \subset V(s) \text{ and } |x|_s \leq |x|_r$$

for x ∈ V(r) if s < r.

We shall identify pairs of functions whose difference is a constant.

If $0 < m < r$ then the elements of the intermediate space V(m) may be characterized by means of the degree of their approximability by means of elements of V(r).

The precise formulation of this fact will be given below; since it is based on some general properties of the scale of spaces V(r) which will be used again later, it will be more convenient to begin with a discussion of these properties.

We shall now adopt a more abstract point of view: keeping in mind the fact that the spaces V(r) are function spaces, the number r being a measure of smoothness, we shall describe them now in a language more suited to formulate and prove the approximation theorems needed in the sequel.

To simplify the notation, we shall adopt the following conventions. We fix a natural number n and denote by T the set of all n-tuples of integers $k = (k_1,\ldots,k_n)$ such that at least one of the k_i is different from zero. For $k \in T$ we define $|k|$ as $(\Sigma\, k_j^2)^{1/2}$. Let V(+∞) be the linear space of all complex valued functions with finite support defined on T. If f, g ∈ V(+∞) and if r is a real number we define

$$[f,g]_r = \Sigma |k|^{2r}\, f(k)\, g(k)^*$$

the sum being extended over the whole of T. For each real r the Hilbert space obtained as the completion of V(+∞) equipped with the scalar product $[\cdot,\cdot]_r$ may be identified with the space V(r) introduced before.

If $r > s$ the space V(r) may be considered in a natural manner as a subspace of V(s); the union of all the V(r) will be denoted by V(-∞). If $x \in V(r)$ and $r > s$ then $x \in V(s)$ as well and $|x|_s \leq |x|_r$. The embedding of V(r) into V(s) is thus seen to be continuous - it is more than continuous though, as we shall see presently.

Given a natural number m we define a projection operator from V(-∞) into V(+∞) as follows. Given f ∈ V(-∞), let $P_m f$ be the element g ∈ V(+∞) such that $g(k) = f(k)$ for $|k| \leq m$ and $g(k) = 0$ otherwise.

11.1 <u>Lemma</u> *If* $r > s$ *and if* P_m *is considered as an operator from* V(r) *into* V(s), *then*

$$|I - P_m| \leq (m + 1)^{s-r}.$$

Proof If $x \in V(r)$ we have

$$|(I - P_m)x|_s^2 = \sum_{|k|>m} |k|^{2s} |x(k)|^2$$

$$= \sum |k|^{2(s-r)} |k|^{2r} |x(k)|^2$$

$$\leq (m+1)^{2(s-r)} \sum_{|k|>m} |k|^{2r}|x(k)|^2 \leq (m+1)^{2(s-r)} |x|_r^2. \quad \square$$

This estimate may be used to prove that the injection of $V(r)$ into $V(s)$ is compact.

11.2 Proposition *If $r > s$ then the natural injection operator $V(r) \rightarrow V(s)$ is compact.*

Proof Consider the space $B(V(r), V(s))$ of all bounded linear operators of $V(r)$ into $V(s)$. If $J(r,s)$ stands for the injection $V(r) \rightarrow V(s)$ we have

$$|J(r,s) - P_m| \leq (m + 1)^{s-r}$$

in the norm of $B(V(r), V(s))$. The injection $J(r,s)$ appears thus as a uniform limit of the finite-rank operators P_m. $\square$

   Now we are ready to state the first of the two lemmata which describe the connection between the approximability of an element by smooth functions and the index of the space in the scale to which it belongs.

11.3 Lemma *Let $0 < m < r$ and suppose $v \in V(m)$ is given. For each $Q \geq 1$ there exists a $w \in V(r)$ such that*

$$|v - w|_0 \leq |v|_m \, Q^{-t}$$

$$|w|_r \leq |v|_m \, Q$$

where $t = \dfrac{m}{r-m}$ or, in other words,

$$\frac{m}{r} = \frac{t}{t + 1} \, .$$

<u>Proof</u> Let N be a fixed natural number and set $w = P_N v$. Considering P_N as an operator from $V(m)$ into $V(0)$, we obtain from (11.1)

$$|v - w|_0 = |(I - P_N)v|_0 \leq (N + 1)^{-m} |v|_m .$$

At the same time,

$$|w|_r^2 = \sum_{|k| \leq N} |k|^{2r} |v(k)|^2 \leq \sum_{|k| \leq N} N^{2(r-m)} |k|^{2m} |v(k)|^2$$

$$= N^{2(r-m)} |v|_m^2$$

so that

$$|w|_r \leq N^{r-m} |v|_m.$$

Hence the lemma will be established if, for each given $Q \geq 1$, a natural number N may be found for which

$$(N + 1)^{-m} \leq Q^{-t}$$

$$N^{r-m} \leq Q.$$

These requirements are equivalent to

$$Q^{\frac{1}{r-m}} - 1 \leq N \leq Q^{\frac{1}{r-m}}$$

so that such a number N exists. The proof is complete. $\quad\square$

11.4 <u>Definition</u> *Consider a fixed subinterval* S *of the real axis. Suppose we are given for each* s, s $\in$ S, *a Banach space* X(s) *such that*

$$X(s_1) \supset X(s_2) \text{ and } |x|_{s_1} \leq |x|_{s_2}$$

for $s_1 < s_2$. *We shall say that the family is logarithmically convex if* $\log |x|_s$ *is a convex function of* s *for each* x $\in$ X(r) *different from zero, in other words, if*

$$|x|_s \leq |x|_{s_1}^{\lambda_1} |x|_{s_2}^{\lambda_2}$$

for non-negative λ_1 *and* λ_2 *with*

$$\lambda_1 + \lambda_2 = 1 \text{ and } s = \lambda_1 \, s_1 + \lambda_2 \, s_2.$$

It is possible to prove that the spaces $V(r)$ defined above form a logarithmically convex family.

The approximation properties which have been proved for the elements of the concrete spaces $V(r)$ are characteristic for the inclusion $x \in V(m)$ in the following sense.

11.5 <u>Lemma</u> *Suppose* $X(s)$, $0 \leq s \leq r$ *is a logarithmically convex family of Banach spaces. Suppose* $v \in X(0)$ *and a number* K *are given.*

Suppose that, for every $Q \geq 1$, *there exists a* $w \in X(r)$ *such that*

$$|v - w|_0 \leq KQ^{-t}$$

$$|w|_r \leq KQ$$

then $v \in X(m)$ *for every* m *satisfying*

$$\frac{m}{r} < \frac{t}{t + 1}.$$

Furthermore, there exists a $c = c(m,t)$ *such that*

$$|v|_m \leq cK.$$

<u>Proof</u> Observe first that, for any s between 0 and r,

$$s = \frac{r - s}{r} \cdot 0 + \frac{s}{r} \cdot r$$

whence

$$|x|_s \leq |x|_0^{\frac{r-s}{r}} \, |x|_r^{\frac{s}{r}} \tag{1}$$

for every $x \in X(r)$.

Choose a $Q \geq 1$. For $n = 0,1,\dots$ take $w_n \in X(r)$ such that

$$|v - w_n|_0 \leq K(2^n Q)^{-t}$$

$$|w_n|_r \leq K \, 2^n \, Q.$$

160

We shall have then

$$|w_n - w_{n+1}|_0 \leq 2K(2^n Q)^{-t}$$

$$|w_n - w_{n+1}|_r \leq K(2^n + 2^{n+1})Q = 3K2^n Q$$

so that, by (1), for any m between Q and r,

$$|w_n - w_{n+1}|_m \leq 3K(2^n Q)^{-t\frac{r-m}{r} + \frac{m}{r}}.$$

Observe that the number

$$q = t\frac{r-m}{r} - \frac{m}{r}$$

will be positive if and only if $\frac{m}{r} < \frac{t}{t+1}$.

Assuming this, we have

$$|w_n - w_{n+1}|_m \leq 3K(2^n Q)^{-q}$$

so that the sequence w_n converges in $X(m)$ to a limit w_∞. Let us show now that $v \in X(m)$.

In order to prove the inclusion $v \in X(m)$ it will be sufficient to show that $v = w_\infty$. This, however, follows from the inequalities

$$|v - w_\infty|_0 = |v - w_n|_0 + |w_n - w_\infty|_0 \leq K(2^n Q)^{-t} + |w_n - w_\infty|_m \rightarrow 0.$$

To obtain an esimate for $|v|_m$, we write

$$|v|_m \leq |v - w_0|_m + |w_0|_m.$$

For the first summand, we obtain

$$|v - w_0|_m = |w_0 - w_\infty|_m \leq \sum_0^\infty |w_n - w_{n+1}|_m \leq 3KQ^{-q} \sum_0^\infty 2^{-qn} = KQ^{-q}C(q)$$

where $C(q) = 3 \sum_0^\infty 2^{-nq} = 3/(1 - 2^{-q})$.

The second summand may be established as follows. First of all, we estimated $|v|_0$. Choosing $Q = 1$, it follows from our assumption that there exists a $w_* \in X(r)$ such that

$$|v - w_*|_0 \leq K$$

$$|v_*|_r \leq K.$$

Hence

$$|v|_0 \leq |v - w_*|_0 + |w_*|_0 \leq |v - w_*|_0 + |w_*|_r \leq 2K$$

and

$$|w_0|_0 \leq |v|_0 + |v - w_0|_0 \leq 2K + KQ^{-t} \leq 3K.$$

It follows that

$$|w_0|_m \leq (3K)^{\frac{r-m}{r}} (KQ)^{\frac{m}{r}} \leq 3K \; Q^{\frac{m}{r}}$$

and

$$|v|_m \leq |v - w_0|_m + |w_0|_m \leq KQ^{-q} \, C(q) + 3KQ^{\frac{m}{r}}.$$

In particular, choosing Q = 1, we can take

$$c(m,t) = C(q) + 3. \qquad\qquad\qquad \square$$

Suppose now we are given two positive numbers r and s and two families of Banach spaces depending on parameters p and q

$$V(r) \subset V(p) \subset V(0)$$

$$G(s) \subset G(q) \subset G(0)$$

if $0 < p < r$ and $0 < q < s$. We shall also assume that the norms satisfy the inequalities

$$|x|_0 \leq |x|_p \leq |x|_r$$

$$|y|_0 \leq |y|_q \leq |y|_s$$

for $x \in V(r)$ and $y \in G(s)$.

Suppose a continuous linear mapping L of V(r) into G(s) is given. Suppose a $g \in G(s)$ is given.

11.6 <u>Definition</u> *Suppose* $g \in G(s)$ *satisfies* $|g|_0 \leq 1.$

We shall say that the equation

$$Lx = g$$

possesses an approximate solution of order t *if there exists a constant* K *with the following property: if* $|g|_s \leq K$ *then, for every* $Q \geq 1$, *there exists a* $w \in V(r)$ *such that*

$$|Lw - g|_0 \leq KQ^{-t},$$

$$|w|_r \leq KQ.$$

In the particular case $G(s) = V(s)$ and L the canonical injection mapping from $V(r)$ into $V(s)$, Lemma 11.3 may be interpreted as the statement that an approximate solution of order $\frac{s}{r-s}$ exists.

Let us turn now to the existence of approximate solutions for a general linear operator L. We shall see that a construction of approximate solutions may be effected using the so-called Galerkin method. This construction actually works in some fairly general classes of Sobolev spaces, we shall restrict ourselves, though, to our concrete scale of Hilbert spaces $V(\cdot)$; this will enable us to put into evidence the properties of the scale which make this construction work and make the meaning of some of the assumptions of the main theorem more plausible.

In order to do so, let us continue our discussion on the spaces $V(\cdot)$. Let us introduce, for each real p, an operator $M(p)$; we define it first on $V(+\infty)$ by the following formula:

$$y = M(p)x$$

is taken to mean

$$y(k) = |k|^p x(k)$$

for each $k \in T$. It is easy to see that, for each real s, the mapping $M(p)$ may be extended to a bounded linear operator $V(s) \to V(s-p)$; it will be convenient to use the name $M(p)$ for any of these extensions. If x and y are elements of $V(s)$ and if $p > 0$ then $M(2p)y \in V(s-p)$ and (since $x \in V(s-p)$ as well)

$$[x, M(2p)y]_{s-p} = \Sigma |k|^{2(s-p)} x(k) |k|^{2p} y(k)^* = \Sigma |k|^{2s} x(k) y(k)^* = [x,y]_s.$$

It is not difficult to see that $M(2p)$ establishes a one-to-one correspondence between the dual of $V(s)$ and $V(s-p)$ based on the equality

$$[x,y]_s = [x,M(2p)y]_{s-p}.$$

In particular, the dual of $V(r)$ may be identified with $V(-r)$, the correspondence being implemented by $M(2r)$, the r-th power of the Laplacian.

To mention another consequence of these facts, let us note that $M(p)$ is an isometric mapping of $V(s)$ onto $V(s-p)$. Indeed, if $x \in V(s)$ then

$$M(p)x \ _{s-p}^2 = [M(p)x, M(p)x]_{s-p} = [x,M(2p)x]_{s-p} = [x,x]_s.$$

Now consider two real numbers r and s, $r > s$ and the embedding $J(r,s)$ of $V(r)$ into $V(s)$. We have seen already that $J(r,s)$ is a compact operator. Let us compute its adjoint: if $x \in V(r)$ and $y \in V(s)$, we have

$$[Jx,y]_s = [x,J^*y]_r$$

and, at the same time,

$$[Jx,y]_s = [x,M(2(s-r))y]_r.$$

It follows that $J(r,s)^* = M(2(s-r))$; in particular $J(r,s)^* J(r,s)$ is nothing more than $M(2(s-r))$ considered as a mapping of $V(r)$ into itself. We shall denote it by $T(r,s)$. The operator $T(r,s)$ is self-adjoint and, as a product of two operators one of which is compact, $T(r,s)$ is compact as well.

To obtain the spectral decomposition of $T(r,s)$ consider a non-negative λ and a nonzero $x \in V(r)$ for which

$$(T(r,s) - \lambda)x = 0.$$

It follows that, for each $k \in T$

$$(|k|^{2(s-r)} - \lambda)x(k) = 0.$$

Hence there exists exactly one $k \in T$ for which $x(k)$ is different from zero and $\lambda = |k|^{2(s-r)}$. For brevity, let us denote, for each $k \in T$, by $e(k)$ the element of $V(-\infty)$ defined as follows:

$$e(k)(t) = 1 \text{ if } t = k$$
$$= 0 \text{ otherwise.}$$

We have thus, for $r > s$,

$$T(r,s) \, e(k) = \frac{1}{|k|^{2(r-s)}} \, e(k). \tag{*}$$

Now let $N > 1$ be given and consider the spectral projection $E(r,s)$ corresponding to the operator $T(r,s)$ and the set

$$\{\lambda; \, |\lambda| \geq N^{2(s-r)}\}.$$

It follows from (*) that $e(k) \in E(r,s)$ if $|k|^{2(s-r)} \geq N^{2(s-r)}$ or, in other words, if and only if $|k| \leq N$. Now define an operator $P(N)$ on $V(-\infty)$ as follows

$$P(N)f = \sum_{|k| \leq N} f(k) \, e(k).$$

Clearly $P(N)$ may be extended to an element of $B(V(r))$ for each r. Let us show now that $P(N)$ is a self-adjoint operator in each $V(r)$. Having shown that, we shall have the equality

$$P(N) = E(r,s)$$

for each r, s with $r > s$. In particular

$$|P(N)|_r = 1$$

for every r.

If $e(k)$ is considered as an element of $V(r)$, we have $|e(k)|_r = |k|^r$; thus $|k|^{-r} e(k)$, $k \in T$ is a complete orthonormal system in $V(r)$. Since $[f, \, |k|^{-r} e(k)]_r = |k|^r f(k)$, we have

$$f(k) = [f, \, |k|^{-r} e(k)]_r |k|^{-r} e(k)$$

so that $P(N)f$ is nothing more than the orthogonal projection in $V(r)$ of f onto the linear span of the $|k|^{-r} e(k)$ for $|k| \leq N$. Our assertion is thus established.

This fact will be essentially used in the proof of the following proposition.

11.7 <u>Proposition</u> *Let* L *be a bounded linear operator from* $V(r)$ *into* $V(s)$, *where r s 0, and let an element* $g \in V(s)$ *be given such that* $|g|_0 \leq 1$. *Suppose the following two conditions are satisfied:*

$$|v|_0^2 \leq (v, Lv)_0 \text{ for } v \in V(r), \qquad (2)$$

$$|v|_s^2 \leq c((v, Lv)_s + K_1^2 |v|_0^2). \qquad (3)$$

Then the equation $Lx = g$ *has an approximate solution of order* $\dfrac{2s-r}{r-s}$.

Proof We intend to construct an approximate solution for $Lx = g$ as the solution of the equation

$$P(Lw - g) = 0 \qquad (4)$$

under the hypothesis

$$Pw = w \qquad (5)$$

where $P = P(N)$ is the spectral projection corresponding to the self-adjoint operator $T(r,s) \in B(V(r))$ and the set $\{\lambda : |\lambda| \geq N^{2(s-r)}\}$, introduced before the statement of our proposition.

In order to show that w exists and is uniquely determined by these two requirements it suffices to show - the system being finite-dimensional - that the only solution of the homogeneous system is zero. This, however follows from the fact that for each $v \in V(r)$ with $P(Lv-g) = 0$ and $Pv = v$ we have

$$|v|_0^2 \leq (v, Lv)_0 = (Pv, Lv)_0 = (Pv, PLv)_0$$

$$= (Pv, Pg)_0 = (v, Pg)_0 \leq |v|_0 |Pg|_0,$$

so that $|v|_0 \leq |Pg|_0$.

It remains to estimate $|v|_r$ and $|Lv - g|_0$. According to the definition of $T(r,s)$ we have

$$(v, w)_s = (v, T(r,s)w)_r, \quad v, w \in V(r)$$

so that

$$|v|_s \geq N^{-(r-s)}|v|_r \text{ if } v \in V(r), \ Pv = v.$$

It follows that

$$|Pv|_r \leq N^{r-s}|Pv|_s \leq N^{r-s}|v|_s, \text{ for } v \in V(r). \tag{6}$$

Now let us consider the self-adjoint operator $T(s,0) \in B(V(s))$ and its spectral projection corresponding to the set $\{\lambda : |\lambda| \geq N^{-2s}\}$. It turns out that this projection equals P when restricted to $V(r)$; we have thus

$$|(I-P)v|_0 \leq N^{-s}|(I-P)v|_s \leq N|v|_s \text{ for } v \in V(r). \tag{7}$$

I. Using (6) we obtain

$$|v|_r^2 = |Pv|_r^2 \leq N^{2(r-s)}\ |v|_s^2$$

$$\leq N^{2(r-s)}\ c\{(v,Lv)_s + K_1^2\ |v|_0^2\} \leq N^{2(r-s)}\ c\{(v,Lv)_s + K_1^2\}$$

since $|v|_0 \leq |Pg|_0 \leq |g|_0 \leq 1$.

Now $(v,Lv)_s = (Pv,Lv)_s = (v,PLv)_s = (v,Pg)_s \leq |v|_s\ |Pg|_s \leq |v|_s\ |g|_s.$

If we set $K = \max\ \{|g|_s, K_1\}$, the estimate above simplifies to

$$|v|_r^2 \leq N^{2(r-s)}\ c(|v|_s\ K + K^2).$$

To estimate $|v|_s$ we observe that

$$|v|_s^2 \leq |v|_r^2 \leq c(|v|_s\ K + K^2)$$

whence $|v|_s \leq (c + 1)K$ so that

$$|v|_r^2 \leq N^{2(r-s)}\ cK(|v|_s + K) \leq N^{2(r-s)}cK(c + 2)K.$$

Thus $|v|_r \leq N^{r-s}\ c_1 K.$

II. Since $P(Lv - g) = 0$ we have

$$|Lv-g|_0 \leq N^{-s}|Lv-g|_s \leq N^{-s}(|Lv|_s + |g|_s)$$

$$\leq N^{-s}\ (c_2|v|_r + K) \leq N^{-s}(c_2 c_1 N^{r-s}\ K + K) \leq K\ c_3\ N^{r-2s}.$$

The degree of approximation is thus

$$\frac{2s - r}{r - s}.$$

$\square$

Let us show now that - under an additional hypothesis - the existence of an approximate solution implies the existence of an exact one.

11.8 **Proposition** *Suppose the family* $V(\cdot)$ *is logarithmically convex and suppose the operator* L *satisfies the following additional hypothesis*

$$|v|_0 \le |Lv|_0 \le c|v|_p \ \text{for every } v \in V(r). \tag{8}$$

(In particular this implies that L *may be extended to a continuous mapping of* $V(p)$ *into* $G(0)$.*)*

   If $g \in G(s)$ *satisfies* $|g|_0 \le 1$, *if* $Lx = g$ *admits an approximate solution of order* $t > \frac{\rho}{r-p}$ *and if* $|g|_s \le K$ *then the equation* $Lx = g$ *admits an exact solution* $v \in V(p)$.

Proof Denote by w_n the approximate solution $w_n \in V(r)$ obtained for $Q_n = 2^n$. We have, using the left inequality in (8),

$$|w_n - w_{n+1}|_0 \le |L(w_n - w_{n+1})|_0 = |Lw_n - g - (Lw_{n+1} - g)|_0 \tag{9}$$

$$\le KQ_n^{-t} + KQ_{n+1}^{-t} \le 2KQ_n^{-t};$$

furthermore

$$|w_n - w_{n+1}|_r \le KQ_n + KQ_{n+1} = 3KQ_n. \tag{10}$$

Suppose now $0 < m < r$; we have the following estimate

$$|w_n - w_{n+1}|_m \le |w_n - w_{n+1}|_r^{\frac{m}{r}} \, |w_n - w_{n+1}|_0^{1-\frac{m}{r}}$$

$$\le (3KQ_n)^{\frac{m}{r}} (2KQ_n^{-t})^{(1 - \frac{m}{r})} \le 3KQ_n^{-q} \tag{11}$$

with $q = t(1 - \frac{m}{r}) - \frac{m}{r}$.

   If q is positive we may conclude that w_n is a Cauchy sequence in $V(m)$; we have thus the existence of an element $v \in V(m)$ such that $v = \lim w_n$ in $V(m)$.

Now q will be positive if and only if $t > \frac{m}{r-m}$. If we choose $m = p$ we shall have q positive by our assumption. At the same time it will follow from the second inequality in (8) that

$$|Lv - g|_0 \leq |L(v - w_n)|_0 + |Lw_n - g|_0$$

$$\leq \sum_{j \geq n} |L(w_{j+1} - w_j)|_0 + |Lw_n - g|_0 \tag{12}$$

$$\leq 3cK(Q_n^{-q} + Q_{n+1}^{-q} + \ldots) + KQ_n^{-t}$$

so that $Lv = g$. □

Now we are ready to state and prove the main theorem which contains sufficient conditions for the existence of the solution of a nonlinear equation. First of all, let us describe the general framework.

Let $(E, |\cdot|)$ be a Banach space and $(E_0, |\cdot|_0)$, $(E_1, |\cdot|_1)$ be two normed vector spaces over the complex field, such that the following conditions hold

1. $E_1 \subset E \subset E_0$

2. There exist numbers $c > 0$ and $0 < \sigma < 1$ such that

$$|u| \leq c|u|_0^{1-\sigma} |u|_1^\sigma$$

for all $u \in E_1$

3. Let R be a positive number and set

$$D = \{u \in E; \ |u| \leq R\}. \tag{13}$$

Let f be a continuous mapping of D into a normed vector space (over the complex field) $(F_0, |\cdot|_0)$. Consider another normed vector space $(F_1, |\cdot|_1)$ such that $F_1 \subset F_0$ and suppose

$$f(D \cap E_1) \subset F_1.$$

We shall make the following assumption about f:

4. (growth) There exist a number S such that

$$|f(u)|_1 \leq \max (S, |u|_1) \text{ for all } u \in D \cap E_1. \tag{14}$$

5. (approximation by a differential) There exists a mapping

$g:(D \cap E_1) \times E_1 \to F_1$, a positive number M and a number β, $0 < \beta < 1$, such that

$$|f(u + v) - f(u) - g(u,v)|_0 = M|v|_0^{2-\beta} |v|_1^{\beta}. \tag{15}$$

6. (solvability of the linearized equation) There exists two positive numbers λ and μ with the following properties: if $u \in D \cap E_1$ and $h \in f(D \cap E_1)$ satisfy $|h|_0 \leq m^{-\lambda}$, where $m = \max (|h|_1, |u|_1)$, then, for each number $Q > 1$, there exists a $v \in E_1$ for which

$$|g(u,v) - h|_0 \leq m Q^{-\mu} \tag{16}$$

$$|v|_1 \leq m Q \tag{17}$$

$$|v|_0 \leq M|g(u,v)|_0. \tag{18}$$

7. The numbers introduced above satisfy the condition

$$0 < \frac{\mu+1}{\mu-\lambda} < \min \left(2-\beta \; \frac{\lambda+(\lambda+1)(\mu+1)}{\lambda(\mu+\beta)} , \; \frac{1-\sigma}{\sigma}\mu - \frac{\mu+1}{\lambda}\right). \tag{19}$$

11.9 Theorem *If conditions 1 - 7 are satisfied then there exists a number $\delta > 0$ such that $|f(0)|_0 < \delta$ implies the existence of an element $u \in (D \cap E_1)^-$ (closure taken in E), for which $f(u) = 0$.*

Proof We shall apply Proposition 1.3 with

$$w(t) = t^a \text{ and } g(t) = t^\omega \tag{20}$$

where a and ω are two positive numbers satisfying the inequalities

$$\frac{\mu+1}{\mu-\lambda} < a < \min \left(2 - \beta \; \frac{\lambda+(\lambda+1)(\mu+1)}{\lambda(\mu+\beta)} , \; \frac{1-\sigma}{\sigma}\mu - \frac{\mu+1}{\lambda}\right), \tag{21}$$

$$1 - \sigma - \sigma \frac{\mu+1}{\mu-\lambda} - \sigma \frac{a}{\mu} > \omega > 0. \tag{22}$$

From (21) it follows in particular that $a > 1$. It is easy to see that if $0 < t < 1$ then the series

$$s(t) = \sum_{n=0}^{\infty} w^n(t) \text{ is convergent}$$

170

moreover if

$$0 < t \le 2^{1/(1-a)} \tag{23}$$

then

$$s(t) \le 2t. \tag{24}$$

To prove the last assertion one has to observe that (23) implies

$$t^{a^n} \le (\tfrac{1}{2})^n t.$$

Let us note that (21) implies

$$\frac{a}{\lambda} > \frac{\mu+1}{\mu\lambda} + \frac{a}{\mu} .$$

Using this observation, together with (22), it follows that there exists an $r_0 > 0$ such that the following inequalities are satisfied for all r between 0 and r_0.

$$r^{2-\beta-\beta} \ (\frac{\mu+1}{\mu\lambda} + \frac{a}{\lambda})^{-a} \ M^{3-\beta_2\beta+\frac{\beta}{\mu}-1} \ 3^{2-\beta} \le 1 \tag{25}$$

$$r^{\frac{a-1}{\lambda}} + r^{\frac{a}{\lambda} - (\frac{\mu+1}{\mu\lambda} + \frac{a}{\mu})} \le 1 \tag{26}$$

$$r^{(1-\sigma-\sigma\frac{\mu+1}{\mu\lambda} -\sigma\frac{a}{\mu})-\omega} \ C(\tfrac{3}{2} M)^{1-\sigma} \ 2^{\frac{\sigma}{\mu}} \le 1 \tag{27}$$

$$rs^{\lambda} \le 1. \tag{28}$$

Let T denote the half open interval $(0,r_0)$ and define for all $r \in T$

$$M(r) = \{u \in E_1; \ |u| \le R-s(r^{\omega}), \ |u|_1 \le r^{-1/\lambda}, \ |f(u)|_0 \le r\}.$$

If we can show that

$$M(r) \subset U(M(r^a), \ r^{\omega}) \tag{29}$$

then according to Proposition 1.3 it will follow that

$$M(t) \subset U(M(0), \ s_g(t)) \tag{30}$$

for all $t \in T$.

Let us take

$$\delta = \min \ (r_0, \ (\tfrac{1}{2})^{1/\omega} \ (a-1), \ (\tfrac{1}{2} R)^{1/\omega}).$$

We have $\delta^\omega \leq (\tfrac{1}{2})^{1/a-1}$ and $2\delta^\omega \leq R$.

From (24) it follows then that $s(\delta^\omega) \leq R$. Thus if we impose the condition $|f(0)| \leq \delta$ we shall have the inclusion $0 \in M(\delta)$. In this case (30) assures the fact that $M(0)$ is nonvoid. But $u \in M(0)$ implies $f(u) = 0$.

Hence the proof of our theorem will be complete if we prove the inclusion (29). Let r be an arbitrary number from T and let $u \in M(r)$.

We intend to show that there exists an element $u' \in M(r^a)$ such that $|u - u'| < r^\omega$. Denote $h = f(u)$ and $m = \max \ (|h|_1, |u|_1)$. Let us prove that $|h|_0 \leq m^{-\lambda}$. Since $|f(u)|_0 \leq r$ it suffices to show that $r \leq m^{-\lambda}$, or in other words

$$r \leq \min \ (|h|_1^{-\lambda}, |u|_1^{-\lambda}).$$

If $|h|_1 \leq |u|_1$ then the above inequality is a simple consequence of the definition of $M(r)$ since $|u|_1 \leq r^{-\lambda}$.

If $|h|_1 \geq |u|_1$, then from the growth condition 4, we have $|h|_1 \leq S$ so that, according to (28), $r \leq S^{-\lambda} \leq |h|_1^{-\lambda}$.

Now let v be the element corresponding to the pair u,h (see 6.) and set $u' = u + v$. Choose Q in such a manner that $mQ^{-\mu} = \tfrac{1}{2} r^a$. From (26) it follows that $r^{a-1} < 1$ so that $\tfrac{1}{2} r^a \leq \tfrac{1}{2} r$. Accordingly to 6. we have

$$|v|_1 \leq mQ = 2^{1/\mu} m^{1+1/\mu} r^{-a/\mu} \leq 2^{1/\mu} r^{-\frac{\mu+1}{\mu\lambda} - \frac{a}{\mu}} \tag{31}$$

$$|v|_0 \leq M|g(u,v)|_0 \leq M(|g(u,v) - h|_0 + |h|_0) \tag{32}$$

$$\leq M(\tfrac{1}{2} r^a + r) = \tfrac{3}{2} Mr.$$

Using these inequalities together with 5. and (25) we obtain

$$|f(u+v)|_0 \leq |f(u+v) - f(u) - g(u,v)|_0 + |g(u,v) - f(u)|_0$$

$$\leq M\, v_0^{2-\beta}\, |v|_1^{\beta} + t \leq M(\tfrac{3}{2} M\, r)^{2-\beta}\, 2^{\frac{\beta}{\mu}}\, r^{-\beta(\frac{\mu+1}{\mu\lambda} + \frac{a}{\mu})} + \tfrac{1}{2}\, r^a$$

$$= M^{3-\beta}\, 2^{\beta + \frac{\beta}{\mu} - 2}\, 3^{2-\beta}\, r^{2-\beta-\beta(\frac{\mu+1}{\mu\lambda} + \frac{a}{\lambda})} + \tfrac{1}{2}\, r^a \leq r^a.$$

Thus we have proved that

$$|f(u')|_0 \leq r^a \tag{33}$$

Now, using (31) and the fact that

$$|u|_1 \leq m \leq r^{-1/\lambda}$$

$$|u + v|_1 \leq |u|_1 + |v|_1 \leq r^{-1/\lambda} + r^{-\frac{\mu+1}{\mu\lambda} - \frac{a}{\mu}}$$

and by virtue of (26) we obtain

$$|u'|_1 \leq r^{-a/\lambda}. \tag{34}$$

From the "interpolating property" 2. it follows that

$$|v| = C|v|_0^{1-\sigma}|v|_1^{\sigma} \leq C(\tfrac{3}{2} M)^{1-\sigma}\, 2^{\frac{\sigma}{r}}\, r^{1-\sigma-\sigma\frac{\mu+1}{\mu\lambda} - \sigma\frac{a}{\mu}}.$$

Using condition (22) we have

$$|u' - u| = |v| \leq r^{\omega}. \tag{35}$$

As an immediate consequence we obtain

$$|u'| \leq |u| + |u' - u| \leq R - s(r^{\omega}) + r^{\omega} = R - s(r^{a\omega}). \tag{36}$$

The relation (33) - (36) show that $u' \in U(M(r^a),\ r^{\omega})$. The proof is complete.   □

The authors would like to thank M. Krbec, M. Štědrý and D. Timotin for helpful discussions about the subject of this section.

12 Appendix A: On the successive iterates of a multidimensional rate of convergence

The aim of this short appendix is to help the reader to understand how the successive iterates of a rate of convergence of type (p.m) are generated. Suppose we have a function

$$w: T^p \to T^m \tag{1}$$

where T can now be an arbitrary set. If $p = m$ then we can iterate the function w in the sense of usual function composition. We obtain the iterates:

$$w^{(0)}(t) = t, \quad w^{(n+1)}(t) = w(w^{(n)}(t)), \quad t \in T^p, \quad n = 0,1,2,\dots . \tag{2}$$

In the case $p \neq m$ the idea is to attach to the function (1) a function

$$\tilde{w}: T^p \to T^p \tag{3}$$

which can be iterated via the usual function composition and then to define the iterates of w by

$$w^{(n)}(t) = w(\tilde{w}^{(n-1)}(t)), \quad t \in T^p, \quad n = 1,2,3,\dots . \tag{4}$$

Let us denote by $w_1, w_2, \dots, w_m$ the components of w, i.e.:

$$w(t) = (w_1(t), \dots, w_m(t)), \quad t \in T^p. \tag{5}$$

Setting

$$w_k^{(n)}(t) = w_k(\tilde{w}^{(n-1)}(t)), \quad t \in T^p \tag{6}$$

we have obviously

$$w^{(n)}(t) = (w_1^{(n)}(t), \dots, w_m^{(n)}(t)), \quad t \in T^p, \quad n \in N, \tag{7}$$

The way of attaching the "square" function $\tilde{w}$ to the function w is not unique. In Section 3 we have chosen a way which was suitable for generalizing the

174

induction theorem. Although we have treated the cases p > m and p < m to-gether there is a clear distinction between them. If p < m then we have a "surplus" of component functions and we can take for $\tilde{w}$ the mapping having as components the last p components of w i.e.:

$$\tilde{w}(t) = (w_{m-p+1}(t),\ldots,w_m(t)) \quad t \in T^p. \tag{8}$$

If p > m then we have a "deficit" of component functions and we must introduce auxiliary functions.

For our purposes it turned out to be convenient to set for any $t = (t_1,t_2,\ldots,t_p) \in T^p$

$$\tilde{w}(t) = (t_{m+1},\ldots,t_p,w_1(t),\ldots,w_m(t)) \tag{9}$$

The iterates $w^{(n)}$ of w can be now obtained as indicated in (4). For example if m = 1 then $w = w_1$ and we have:

$$w^{(2)}(t) = w(t_2,\ldots,t_p,w(t))$$

$$w^{(3)}(t) = w(t_3,\ldots,t_p,w(t),w^{(2)}(t))$$

. . . .

$$w^{(p+1)}(t) = w(w(t),w^{(2)}(t),\ldots,w^{(p)}(t))$$

. . . .

We can formally handle the cases p > m and p $\leq$ m together by setting for any $t = (t_1,\ldots,t_p)$

$$w_{-p+1}(t) = t_1, \; w_{-p+2}(t) = t_2,\ldots,w_0(t) = t_p. \tag{10}$$

With this notation (9) reduces formally to (8). Formula (6) can be used for defining $w_k^{(n)}$ of all functions w_k with -p+1 $\leq$ k $\leq$ m. We have

$$\tilde{w}^{(n)} = (w_{m-p+1}^{(n)},\ldots,w_m^{(n)}) \tag{11}$$

for p > m as well as for p $\leq$ m.

The following two diagrams illustrate how the successive iterates of $\tilde{w}$ are generated. To clarify the situation we have written t_k instead of the function w_{-p+k}.

The function in the n-th row and k-th column of the diagram is $w_k^{(n)}$.

__Diagram 1__ : $p = 3$, $m = 5$, $m + 1 - p = 3$, $\tilde{w} = (w_3\ w_4\ w_5)$

	-2	-1	0	1	2	3	4	5
0						t_1	t_2	t_3
1	t_1	t_2	t_3	w_1	w_2	w_3	w_4	w_5
2	w_3	w_4	w_5	$w_1(\tilde{w})$	$w_2(\tilde{w})$	$w_3(\tilde{w})$	$w_4(\tilde{w})$	$w_5(\tilde{w})$
3	$w_3(w)$	$w_4(w)$	$w_5(w)$	$w_1(\tilde{w}^{(2)})$	$w_2(\tilde{w}^{(2)})$	$w_3(\tilde{w}^{(2)})$	$w_4(\tilde{w}^{(2)})$	$w_5(\tilde{w}^{(2)})$

Diagram 2 : $p = 5$, $m = 3$, $m + 1 - p = -1$, $\tilde{w} = (t_4\ t_5\ w_1\ w_2\ w_3)$

	-4	-3	-2	-1	0	1	2	3
0				t_1	t_2	t_3	t_4	t_5
1	t_1	t_2	t_3	t_4	t_5	w_1	w_2	w_3
2	t_4	t_5	w_1	w_2	w_3	$w_1(\tilde{w})$	$w_2(\tilde{w})$	$w_3(\tilde{w})$
3	w_2	w_3	$w_1(\tilde{w})$	$w_2(\tilde{w})$	$w_3(\tilde{w})$	$w_1(\tilde{w}^{(2)})$	$w_2(\tilde{w}^{(2)})$	$w_3(\tilde{w}^{(2)})$

13 Appendix B: Fréchet derivatives, divided differences and consistent approximations

Fréchet differentiability

Let f be a mapping (nonlinear in general) defined on an open subset D of a Banach space X and with values in a Banach space Y. Consider an $x_0 \in D$.

13.1 __Definition__ *We say that f is Fréchet differentiable at the point* x_0 *if there exists a linear operator* $A \in B(X,Y)$ *such that*

$$\lim_{|h| \to 0} \frac{1}{|h|} |f(x_0 + h) - f(x_0) - A h| = 0.$$

It is easy to see that such an operator is unique if it exists. This operator A will be called the Fréchet derivative of f at the point x_0 and will be denoted by $f'(x_0)$.

The following facts may be proved using methods analogous to those of real analysis:

1. If f is Fréchet differentiable at the point x_0 then f is continuous at this point.

2. If $S \in B(X,Y)$ and if $f(x) = Sx$ for all $x \in X$, then f is Fréchet differentiable at every point $x \in X$ and $f'(x) = S$.

3. Let f and g be two mappings defined on D and taking values in Y. If f and g are Fréchet differentiable at a point $x_0 \in D$, then the mapping f + g is also Fréchet differentiable at x_0 and $(f + g)'(x_0) = f'(x_0) + g'(x_0)$.

4. Let X,Y and Z be three Banach spaces and consider the mappings

$$f : D_f \subset X \to Y, \quad g : D_g \subset Y \to Z.$$

Let x_0 be an interior point of D_f such that $y_0 = f(x_0)$ is an interior point of D_g. If f is Fréchet differentiable at x_0 and g is Fréchet differentiable at y_0 then the mapping $h = g \circ f$ is Fréchet differentiable at x_0 and $h'(x_0) = g'(y_0) f'(x_0)$.

If the mapping f is Fréchet differentiable at every point of D, then we

shall say that f is Fréchet differentiable on D. In this case we may consider a mapping f':D → B(X,Y) which associates to each point x ∈ D the Fréchet derivative of f at x. This mapping will be called the Fréchet derivative
of f.

Let x,y be two points from D and suppose that the segment [x,y] = {x+t(y-x);
t ∈ [0,1]} is contained in D. Consider a continuous linear functional
y'∈ ^{d}Y. Denote h = y - x and consider the real function

$$\phi(t) = \langle f(x + t\,h), y'\rangle.$$

If f is Fréchet differentiable at each point of the segment [x,y] then it will
follow from 2. and 4. that φ is differentiable on the interval [0,1] and

$$\phi'(t) = \langle f'(x + th)h, y'\rangle.$$

Suppose now that

$$\mu = \sup_{t\in[0,1]} |f'(x + t(y - x))| < \infty \ ;$$

then we have

$$|\langle f(y) - f(x), y'\rangle| = |\phi(1) - \phi(0)|$$

$$\leq \sup_{t\in[0,1]} |\phi'(t)| \leq \mu|y'| \cdot |y - x|.$$

Since $|f(y) - f(x)| = \sup_{|y'|\leq 1} |\langle f(y) - f(x), y'\rangle|$, we infer that

$$|f(y) - f(x)| \leq \sup_{t\in[0,1]} |f'(x + t(y - x)| \cdot |y - x| \qquad (*)$$

In particular we have obtained the following result.

13.2 Proposition *Let D be a convex open subset of a Banach space X and
let f be a mapping from X to a Banach space Y. If f is Fréchet differentiable on D and if there exists a constant M such that*

$$|f'(x)| \leq M \ \textit{for all}\ x \in D,\ \textit{then}\ |f(x)-f(y)| \leq M\ |x-y|\ \textit{for all}\ x,y \in D.$$

Formula (*) is an analogon of the well known mean value formula from real
analysis. If the mapping f' is Riemann integrable on the segment [x,y] we

can give the following integral representation of the mean value formula

$$f(x) - f(y) = (\int_0^1 f'(x + t(y - x))dt)\ (x - y),\qquad(1)$$

which can be proved using the corresponding formula from real analysis, applied to the function ϕ defined above.

Divided differences of an operator

A notion which is closely related to the notion of Fréchet derivative is the notion of divided difference of an operator. This notion was introduced by J. Schröder [63] and it generalizes the usual notion of divided difference of a scalar function in the same way in which the Fréchet derivative generalizes the notion of derivative of a function.

Let f be, as before, a (nonlinear) mapping defined on $D \subset X$ and with values in Y and let x,y be two distinct points of D.

13.3 <u>Definition</u>   *A linear and bounded operator from X to Y, denoted [x,y;f], which satisfies the condition*

$$[x,y;f]\ (x - y) = f(x) - f(y).\qquad(2)$$

will be called a divided difference of f *at the points* x *and* y.

Of course the above condition does not determine uniquely the divided difference, with the exception of the case when X is one-dimensional. It can be proved that if dim X = d and dim Y = d' then there exist d'(d-1) + 1 linear operators from E to F which satisfy (2) and are linearly independent.

Let now D be a *convex* open subset of X and let us suppose that we have associated to each pair (x,y) of distinct points from D a divided difference [x,y;f] of f at these points. In applications one often has to require that the mapping $(x,y) \to [x,y;f]$ satisfy a Lipschitz condition. We shall suppose that there exists a non-negative constant h such that

$$|[x,y;f] - [x',y';f]| \leq h(|x - x'| + |y - y'|)\qquad(3)$$

for all x,y, x', y' $\in$ U with x $\neq$ y and x' $\neq$ y'.

We shall say in this case that f *has a Lipschitz continuous divided difference on* D. This condition allows us to extend by continuity the mapping

$(x,y) \rightarrow [x,y;f]$ to the whole Cartesian product $D \times D$. From (2) and (3) it follows that f is Fréchet differentiable on D and that $[x,x;f] = f'(x)$. It also follows that the Fréchet derivative of f satisfies a Lipschitz condition of the form:

$$|f'(x) - f'(y)| \leq k \, |x - y| \text{ with } k = 2h. \tag{4}$$

It will also be useful to note that

$$|[x,y;f] - f'(z)| \leq h(|x - z| + |y - z|) \tag{3''}$$

for all $x,y,z \in U$.

Conversely if we suppose that f is Fréchet differentiable on D and that its Fréchet derivative satisfies (4) then it follows that f has a Lipschitz continuous divided difference on D. Indeed, we may take for example

$$[x,y;f] = \int_0^1 f'(x + t(y - x))dt. \tag{5}$$

However, with the exception of the case where dim X = 1, this is not the only Lipschitz continuous divided difference of f. In the following proposition we shall give a characterization of the divided differences of the form (5).

13.4 <u>Proposition</u> *Let $[\cdot,\cdot; f]:D \times D \rightarrow B(X,Y)$ be a mapping which satisfies conditions (2) and (3). The following two assertions are equivalent:*

(i) *Equality (5) holds for all $x,y \in D$.*

(ii) *For all points $u,v \in D$ such that $2v - u \in D$ we have*

$$[u,v;f] = 2[u,2v - u;f] - [v,2v - u;f]. \tag{6}$$

<u>Proof</u> Substituting $w = v - u$, we can rewrite (6) in the form

$$[u, u + w;f] + [u + w, u + 2w;f] = 2[u,u + 2w;f] \tag{6'}$$

Then the implication (i) $\Rightarrow$ (ii) follows immediately observing that

$$\int_0^1 f'(u + tw)dt + \int_0^1 f'(u + w + tw)dt$$

180

$$= \lim_{n \to \infty} \frac{1}{n} \left[\sum_{k=1}^{n} f'(u + \frac{k}{n} w) + \sum_{k=1}^{n} f'(u + w + \frac{k}{n} w) \right]$$

$$= 2 \lim_{n \to \infty} \frac{1}{2n} \sum_{k=1}^{2n} f'(u + \frac{k}{2n} 2w) = 2 \int_{0}^{1} f'(u + t2w)dt.$$

In order to prove the implication (ii) ⇒ (i), let us first observe that (6') implies

$$2^n[u, u + 2^n w; f] = \sum_{k=1}^{2^n} [u + (k-1)w, u + kw;f], \qquad (6'')$$

for all $n \in \mathbb{N}$.

This is easily proved by induction. To see this, consider the first four summands. We have

$$[u, u + w;f] + [u+w,u + 2u;f] + [u + 2w,u+3w;f] + [u+3w,u+4w;f]$$

$$= 2[u,u + 2w;f] + 2[u + 2w, u + 4w;f] = 4[u,u + 4w;f]$$

and it is now obvious how the induction goes on.

Considering now the equality (6'') for $u = x$ and $w = 2^{-n}(y-x)$ and using (3'') we obtain the estimate

$$\left| [x,y;f] - 2^{-n} \sum_{k=1}^{2^n} f'(x + kw) \right|$$

$$= \frac{1}{2^n} \left| \sum_{k=1}^{2^n} ([x + (k-1)w_1 x + kw;f] - f'(x + kw)) \right|$$

$$\leq 2^{-n} \cdot 2^n h |w| = \frac{1}{2^n} h |y - x|.$$

Letting n tend to infinity, we obtain relation (5). □

Consistent approximations of the derivative

The Lipschitz condition (3) has been used for examples in [17], [21], [30], [32], [39] and [42]. In proving the convergence of the secant method A. Sergeev [65] and S. Ulm [68] have assumed a Lipschitz condition of the form:

$$|[x,y;f] - [y,z;f]| \leq h |x-z|. \qquad (7)$$

It is easy to see that (7) implies (3). Indeed we have

$$|[x,y;f] - [x',y';f]| \leq |[x,y;f] - [y,x';f]|$$

$$+ |[y,x'f] - [x',y';f]| \leq h|x-x'| + h|y-y'|.$$

On the other hand by taking x = x in (7) it follows that

$$[x,y;f] = [y,x;f].$$

Many important examples of divided differences satisfy condition (3) but they do not satisfy the above symmetry relation and consequently they do not satisfy (7).

J.W. Schmidt [59] has proved the convergence of the secant method assuming the condition

$$|[x,y;f] - [y,z;f]| \leq a|x-z| + b(|x-y| + |y-z|) \tag{3'}$$

which is more general than (3).

As we have already mentioned conditions (2) and (3) imply the Frechet differentiability of f and the fact that $f'(x) = [x,x;f]$. The same thing follows from (2) and (3'). Taking x' = y' = z in (3) we obtain

$$|[x,y;f] - f'(z)| \leq h(|x-z| + |y-z|). \tag{8}$$

This inequality can also be obtained from (3') with h = a + b.

It turned out that the essential role in the proof of the convergence of the secant method is played by the "approximation property" (8) and not by the "interpolation property" (2). (See [6], [60] and [24]).

13.5 **Definition** *Let X and Y be two Banach spaces and let D be a convex open subset of E. Let us consider a nonlinear operator* f;D → Y *which is Fréchet differentiable on D. A mapping* δf;D × D → B(X,Y) *will be called a consistent approximation of* f' *on D if there exists a constant* h *such that*

$$|\delta f(x,y) - f'(z)| \leq h(|x - z| + |y - z|)$$

for all x,y, z ∈ D.

For operators whose derivative satisfies a Lipschitz condition with

182

constant k we had the following estimate for the distance of the operator from its local linear part

$$|f(y) - f(x) - f'(x) (y - x)| \leq \frac{1}{2} k |y - x|^2 \tag{9}$$

If we use a consistent approximation with constant h then f' will be Lipschitz with constant 2h and the above estimate has to be replaced by one of the following

13.6 Proposition *If δf is a consistent approximation of* f' *on* D *then each of the following four expressions is an estimate for*

$$|f(x) - f(y) - \delta f(u,v) (x-y)|$$

$$e_1 = h(|x-u| + |y-u| + |u-v|) |x-y|$$

$$e_2 = h(|x-v| + |y-v| + |u-v|) |x-y|$$

$$e_3 = h(|x-y| + |y-u| + |y-v|) |x-y|$$

$$e_4 = h(|x-y| + |x-u| + |x-v|) |x-y|.$$

Proof It will be sufficient to give the proof for e_1 and e_3 since the proof for e_2 is obtained from that for e_1 on interchanging u and v and, in a similar manner a proof for e_4 may be obtained by considering x instead of y in appropriate places.

To prove e_1, observe that

$$f(x) - f(y) - f'(u) (x-y) = \int_0^1 (f'(y+t(x-y))-f'(u))dt(x-y)$$

and

$$|f'(y+t(x-y)) - f'(u)| \leq 2h(t|x-u| + (1 - t) |y-u|).$$

It follows that

$$|f(x) - f(y) - f'(u)(x-y)| \leq 2h(\frac{1}{2}|x-u| + \frac{1}{2}|y-u|)|x-y|$$

$$= h(|x-u| + |y-u|) |x-y|.$$

At the same time

183

$$|f'(u) - \delta f(u,v)| \leq h\ |u - v|.$$

Together with the preceding estimate this yields

$$|f(x) - f(y) - \delta f(u,v)(x-y)| \leq h(|x-u| + |y-u| + |u-v|)\ |x-y|$$

so that e_1 is established.

The proof for e_3 is straightforward

$$|f(x) - f(y) - \delta f(u,v)(x-y)| \leq |f(x)-f(y)-f'(y)(x-y)|$$

$$+ |(f'(y) - \delta f(u,v))\ (x-y)|$$

$$\leq h\ |x-y|^2 + h(|y-u| + |y-v|)|x-y|. \qquad \square$$

A divided difference in $\mathbb{R}^d$

In the sequel we shall give an example of a Lipschitz continuous divided difference of an operator acting in a finite dimensional space. We shall consider the space $\mathbb{R}^d$ equipped with the Chebyshev norm, which is given for each $x = (x_1,\ldots,x_d)^T \in \mathbb{R}^d$ by the expression

$$|x| = \max_{1 \leq i \leq d}\ |x_i|.$$

In this case it follows that the norm of a linear operator $A \in B(\mathbb{R}^d)$, represented by the matrix with entries a_{ij}, is

$$|A| = \max_{1 \leq i \leq d}\ \sum_{j=1}^{d}\ |a_{ij}|.$$

Let U be an open sphere of $\mathbb{R}^d$ and let f be an operator defined on U and with values in $\mathbb{R}^d$. Let us denote by $f_1,\ldots,f_d$ the components of f. For each $x \in U$ we shall have

$$f(x) = (f_1(x),\ldots,f_d(x))^T.$$

It is easy to see that the Fréchet derivative of f at a point x will be a d by d matrix having the entry $(\partial f_i)/(\partial x_j)(x)$ in row i and column j, in other words, $f'(x)$ will be the Jacobian of f. In what follows it will be convenient to use the notation

$$D_j f_i = \frac{\partial f_i}{\partial x_j} .$$

Let x,y be two distinct points from U and let us denote by [x,y;f] the matrix having the entries

$$[x,y;f]_{ij} = \frac{1}{x_j - y_j} (f_i(x_1,\ldots,x_{j-1},x_j,y_{j+1},\ldots,y_d)$$

$$- f_i(x_1,\ldots,x_{j-1},y_j,y_{j+1},\ldots,y_d)).$$

The linear operator [x,y;f] defined in this way obviously satisfies condition (2). We shall show that if the partial derivatives $D_j f_i$ satisfy some Lipschitz conditions of the form

$$|D_j f_i(x_1,\ldots,x_k + t,\ldots,x_d) - D_j f_i(x_1,\ldots,x_k,\ldots,x_d)| \leq h_{jk}^i |t| \quad (11)$$

then condition (3) will also be satisfied with

$$h = \max_{1 \leq i \leq d} \{ \frac{1}{2} \sum_{j=1}^{d} h_{jj}^i + \sum_{j=1}^{d} \sum_{k=1}^{j-1} h_{jk}^i \}.$$

It will be sufficient to prove that for all points x,y, u ∈ U with x ≠ y and u ≠ y the following two inequalities are satisfied:

$$|[y,x;f] - [u,x;f]| \leq h|y-u|$$

$$|[y,x;f] - [y,u;f]| \leq h|x-u|.$$

We shall prove only the first inequality, the proof of the second one being analogous.

The proof relies on an identity which will be used twice and which is interesting in its own right.

Given a function h of m variables and given two points $y = (y_1,\ldots,y_m)$ and $u = (u_1,\ldots,u_m)$ in its domain of definition, the difference $h(y) - h(u)$ may be expressed as a sum of differences such that in each of them only one variable is changed. In fact

$$h(y_1,\ldots,y_m) - h(u_1,\ldots,u_m)$$

$$= \sum_{k=1}^{m} (h(u_1,\ldots,u_{k-1},y_k,\ldots,y_m) - h(u_1,\ldots,u_k,y_{k+1},\ldots,y_m)) .$$

To avoid a possible misunderstanding, let us write down the terms for k = 1 and k = m. The first term is

$$h(y_1,y_2,\ldots,y_m) - h(u_1,y_2,\ldots,y_m)$$

and the last one

$$h(u_1,\ldots,u_{m-1},y_m) - h(u_1,\ldots,u_m).$$

To prove the desired estimate, consider a fixed pair i,j. We have

$$[y,x;f]_{ij} - [u,x;f]_{ij}$$

$$= \frac{1}{y_j-x_j} (f_i(y_1,\ldots,y_{j-1},y_j,x_{j+1},\ldots,x_d)-f_i(y_1,\ldots,y_{j-1},x_j,x_{j+1},\ldots,x_d))$$

$$- \frac{1}{u_j-x_j} (f_i(u_1,\ldots,u_{j-1},u_j,x_{j+1},\ldots,x_d)-f_i(u_1,\ldots,u_{j-1},x_j,x_{j+1},\ldots,x_d))$$

$$= \frac{1}{y_j-x_j} (f_i(y_1,\ldots,u_{j-1},y_j,x_{j+1},\ldots,x_d)-f_i(u_1,\ldots,u_{j-1},x_j,x_{j+1},\ldots,x_d))$$

$$- \frac{1}{u_j-x_j} (f_i(u_1,\ldots,u_{j-1},u_j,x_{j+1},\ldots,x_d)-f_i(u_1,\ldots,u_{j-1},x_j,x_{j+1},\ldots,x_d))$$

$$+ \frac{1}{y_j-x_j} (f_i(y_1,\ldots,y_{j-1},y_j,x_{j+1},\ldots,x_d)-f_i(y_j,\ldots,y_{j-1},x_j,x_{j+1},\ldots,x_d))$$

$$- f_i(u_1,\ldots,u_{j-1},y_j,x_{j+1},\ldots,x_d) + f_i(u_1,\ldots,u_{j-1},x_j,x_{j+1},\ldots,x_d) .$$

The sum of the first two terms clearly equals

$$\int_0^1 \{D_j f_i(u_1,\ldots,u_{j-1}, x_j + t(y_j-x_j), x_{j+1},\ldots,x_d)$$

$$- D_j f_i(u_1,\ldots,u_{j-1}, x_j + t(u_j-x_j), x_{j+1},\ldots,x_d)\}dt.$$

The third term will be rewritten as a sum of step by step differences as follows: introducing the abbreviations

$$y(t_1,\ldots,t_{j-1}) = f_i(t_1,\ldots,t_{j-1},y_j,x_{j+1},\ldots,x_d)$$

$$x(t_1,\ldots,t_{j-1}) = f_i(t_1,\ldots,t_{j-1},x_j,x_{j+1},\ldots,x_d)$$

we may rewrite it as

186

$$\frac{1}{y_j - x_j} \{y(y_1, \ldots, y_{j-1}) - y(u_1, \ldots, u_{j-1})$$

$$- (x(y_1, \ldots, y_{j-1}) - x(u_1, \ldots, u_{j-1}))\}$$

$$= \frac{1}{y_j - x_j} \sum_{k=1}^{j-1} (y(u_1, \ldots, u_{k-1}, y_k, \ldots, y_{j-1}) - x(u_1, \ldots, u_{k-1}, y_k, \ldots, y_{j-1})$$

$$+ x(u_1, \ldots, u_k, y_{k+1}, \ldots, y_{j-1}) - y(u_1, \ldots, u_k, y_{k+1}, \ldots, y_{j-1})).$$

Representing the divided differences as integrals of the function $D_j f_i$ in an analogous manner as for the first two terms we obtain

$$[y, x; f] - [u, x; f]$$

$$= \int_0^1 \{D_j f_i(u_1, \ldots, u_{j-1}, x_j + t(y_j - x_j), x_{j+1}, \ldots, x_d)$$

$$- D_j f_i(u_1, \ldots, u_{j-1}, x_j + t(u_j - x_j), x_{j+1}, \ldots, x_d)\}dt$$

$$+ \sum_{k=1}^{j-1} \int_0^1 \{D_j f_i(u_1, \ldots, u_{k-1}\ y_k, \ldots, y_{j-1}, x_j + t(y_j - x_j), x_{j+1}, \ldots, x_d)$$

$$- D_j f_i(u_1, \ldots, u_k, y_{k+1}, \ldots, y_{j-1}, x_j + t(y_j - x_j), x_{j+1}, \ldots, x_d)\}dt.$$

Applying now the estimate (11) we obtain

$$|[y, x; f]_{ij} - [u, x; f]_{ij}| \leq \frac{1}{2}|y_j - u_j| h_{jj}^i + \sum_{k=1}^{j-1} |y_k - u_k| h_{jk}^i$$

$$\leq |y - u| \left(\frac{1}{2} h_{jj}^i + \sum_{k=1}^{j-1} h_{jk}^i\right)$$

whence it follows that

$$[y, x; f] - [u, x; f] \leq h|y - u|.$$

Fréchet derivatives and divided differences of higher order

In the end of this Appendix we want to give a brief exposition of the notions of Fréchet derivative and divided difference of higher order. First we shall introduce the notion of a bounded multilinear operator.

13.7 __Definition__ *Let* X *and* Y *be two Banach spaces and let* n *be a natural number. A mapping* $A:X^n \to Y$ *will be called a bounded* n-*linear operator from* X *to* Y *if the following two conditions are satisfied*

1. *the mapping* $(x_1,\ldots,x_n) \to A(x_1,\ldots,x_n)$ *is linear in each variable* x_i.

2. *there exists a constant* a *such that*

$$|A(x_1,\ldots,x_n)| \leq a|x_1| \cdots |x_n|.$$

The norm of a bounded n-*linear operator can be defined by the formula*

$$|A| = \sup \{|A(x_1,\ldots,x_n)|; \; |x_1| = \ldots = |x_n| = 1\}.$$

Now let B(X,Y) denote, as before, the set of all bounded linear operators from X to Y. Set $B^{(1)}(X,Y) = B(X,Y)$ and define recursively

$$B^{(k+1)}(X,Y) = B(X,B^{(k)}(X,Y)), \; k = 1,2,3,\ldots .$$

In this way one obtains a sequence of Banach spaces $B^{(n)}(X,Y)$, $n = 1,2,\ldots$.

Every $A \in B^{(n)}(X,Y)$ can be viewed as a bounded n-linear operator if one takes

$$A(x_1,\ldots,x_n) = (\ldots((Ax_1)x_2)x_3 \ldots)x_n.$$

In the right hand side of the above equality one has

$$Ax_1 \in B^{(n-1)}(X,Y), \; (Ax_1)x_2 \in B^{(n-2)}(X,Y), \; \text{etc.}$$

Conversely any bounded n-linear operator A from X to Y can be interpreted as an element of the space $B^{(n)}(X,Y)$. Moreover the norm of A as a bounded n-linear operator coincides with the norm of A as an element of the space $B^{(n)}(X,Y)$. Thus we may identify this space with the space of all bounded n-linear operators from X to Y.

To simplify the notations in the sequel we shall write $Ax_1,\ldots,x_n$ for $A(x_1,\ldots,x_n)$. If $x_1 = x_2 = \ldots = x_n$ we shall write Ax^n for $Ax,\ldots,x$.

Now let us consider again a nonlinear operator $f:D \to Y$ where D is an open subset of X. Suppose this operator is Fréchet differentiable on D. Then we may consider the mapping $f':D \to B(X,Y)$ which associates to each point x the Fréchet derivative of f at x. If the mapping f' is Fréchet differentiable at a point $x_0 \in D$ then we shall say that f is twice Fréchet differentiable at

x_0. The Fréchet derivative of f' at x_0 will be denoted by $f''(x_0)$ and will be called the second Fréchet derivative of f at x_0. Of course $f''(x_0)$ will belong to the space $B^{(2)}(X,Y)$.

In a similar way we may define the Fréchet derivatives of higher order. If f is n times Fréchet differentiable at x_0 then we shall call $f^{(n)}(x_0)$ the n-th Fréchet derivative of f at x_0. We shall have obviously $f^{(n)}(x_0) \in B^{(n)}(X,Y)$.

We can also define divided differences of higher order for f. Let us consider a mapping $[\cdot,\cdot;f]$ which associates to each pair $(x,y) \in D \times D$ a linear operator $[x,y;f] \in B(X,Y)$ satisfying (2).

If we fix the first variable, let us say if we put $x = x_0$, we get a mapping

$$[x_0,\cdot;f] : D \to B(X,Y). \tag{12}$$

Let x_1, x_2 be two distinct points of D. A divided difference of the mapping (12) at the points x_1, x_2 will be called a divided difference of the second order of f at the points $x_0; x_1, x_2$. We shall write

$$[x_0,x_1,x_2;f] = [x_1,x_2: [x_0,\cdot;f]]. \tag{13}$$

Generally we shall have

$$[x_0,x_1,\ldots,x_{k-1},x_k,x_{k+1};f] = [x_k,x_{k+1};[x_0,x_1,\ldots,x_{k-1},\cdot;f]] \quad k = 1,2,3,\ldots$$

The n-linear operator $[x_0,x_1,\ldots,x_n;f] \in B^{(n)}(X,Y)$ will be called a divided difference of the n-th order of f at the points $x_0,x_1,\ldots,x_n$.

From (12) it follows that

$$[x_0,\ldots,x_{k-1},x_k,x_{k+1},f] (x_k - x_{k+1})$$

$$= [x_0,\ldots,x_{k-1},x_k;f] - [x_0,\ldots,x_{k-1},x_{k+1};f]. \tag{14}$$

Using the above relation we shall prove that for every $n \in \mathbb{Z}_+$ and for all points y, x_0. $x_1,\ldots,x_n \in D$ the following equality (Newton formula) is satisfied:

$$f(y) = f(x_0) + [x_0,x_1;f] \ (y - x_0) + [x_0,x_1,x_2;f]$$

$$(y - x_1) \ (y - x_0) + \ldots + [x_0,x_1,\ldots,x_n;f] \ (y - x_{n-1}) \ldots$$

$$\ldots (y - x_0) + [x_0,\ldots,x_n,y;f] \ (y - x_n) \ (y - x_{n-1}) \ldots (y - x_0). \quad (15)$$

Denoting $R_n(y) = [x_0,\ldots,x_n,y;f] \ (y - x_n) \ (y - x_{n-1}),\ldots, (y - x_0)$ the above formula may be written as

$$f(y) = L_n(y) + R_n(y). \qquad\qquad (15')$$

Taking formally $[x_0;f] = f(x_0)$ the equality (15) reduces for $n = 0$ to $f(y) - f(x_0) = [x_0,y;f] \ (y - x_0)$. Let us suppose that (15) is verified for a certain $n \geq 0$. Using (14) we have $[x_0,\ldots,x_n,y;f] = [x_0,\ldots,x_n,x_{n+1};f] + [x_0,\ldots,x_n, x_{n+1},y;f] \ (y - x_{n+1})$, whence

$$R_n(y) = [x_0,\ldots,x_n,x_{n+1};f] \ (y-x_n)(y-x_{n-1}),\ldots,(y-x_0) + R_{n+1}(y). \ (16)$$

Thus (15) will be valid for $n + 1$ as well.

Let D be convex and open. If the k-th derivative of f is continuous on D we can take, by analogy with (5),

$$[x_0,\ldots,x_k;f] = \int_0^1 \ldots \int_0^1 t_1^{k-1} \ t_2^{k-2} \ldots t_{k-1} f(x_0 + t_1(x_1-x_0)$$

$$+ \ t_1 t_2(x_2-x_1) + \ldots + t_1 t_2,\ldots,t_k(x_k-x_{k-1}))dt_1 \ dt_2,\ldots,dt_k.$$

It is easy to see that the multilinear operators defined above verify (14), so that they are indeed divided differences of k-th order of f.

In what follows we shall assume that all the divided differences are of the form (16). In this case for $x_0 = x_1 = \ldots = x_k = x$ we shall have

$$\underbrace{[x,x,\ldots,x;f]}_{k+1 \text{ times}} = \frac{1}{k} \ f^{(k)}(x).$$

Suppose now that the n-th Fréchet derivative of f is Lipschitz continuous on D, i.e. there exists a constant k_{n+1} such that

$$|f^{(n)}(u) - f^{(n)}(v)| \leq k_{n+1} \ |u - v| \qquad\qquad (17)$$

for all $u, v \in D$. In this case, writing

$$R_n(y) = ([x_0,\ldots,x_{n-1},y;f] - [x_0,\ldots,x_{n-1},x_n;f])(y-x_{n-1}),\ldots,(y-x_0)$$

and using the representation (16) it is easy to infer that

$$|R_n(y)| \leq \frac{k_{n+1}}{(n+1)!} \, |y - x_n| \, |y - x_{n-1}| \, \cdots \, |y - x_0|.$$

Taking now in (15) $y = x + h$ and $x_0 = x_1 = \ldots = x_n = x$ we deduce the following inequality

$$|f(x+h) - (f(x) + f'(x)h + \frac{1}{2} f''(x)h^2 + \ldots + \frac{1}{n!} f^{(n)}(x)h^n)|$$

$$\leq \frac{k_{n+1}}{(n+1)!} \, |h|^{n+1}$$

which is the analogon of the Taylor formula. In particular, for $n = 1$ this formula reduces to (9).

Let us remark in the end that from Proposition 13.2 it follows that if $f^{(n+1)}$ exists on D and if $|f^{(n+1)}(x)| \leq k_{n+1}$ then (17) is satisfied.

14 Appendix C: Explicit expressions for some finite sums

To obtain estimates for the distance of x_n to the solution, we need explicit expressions for

$$s(w^{(n)}(r)) = s(r) - \sum_{j=0}^{n-1} w^{(j)}(r) \quad n = 0,1,\ldots .$$

Let us now describe how such expressions may be obtained for the rates of convergence considered in Section 2. First we shall study the function

$$w(r) = \frac{r^2}{2(r^2 + a^2)^{1/2}} .$$

The case $a = 0$ presents no difficulties, the function being linear; accordingly, we may limit ourselves to the case $a > 0$. It seems to be difficult to compute directly the iterates $w^{(n)}$, the expressions become complicated; on the other hand, the method suggested in the Gatlinburg lecture [49] may be used to establish explicit formulae for the partial sums

$$x + w(x) + \ldots + w^{(n)}(x).$$

Consider the function $f(x) = x^2 - a^2$ where a is a non-negative number. If $x > a$ define $N(x)$ to be the result of applying one step of the Newton process for f starting at x. It follows that

$$N(x) = x - f'(x)^{-1} f(x) = \frac{1}{2} (x + \frac{a^2}{x})$$

and it is easy to see that $a < N(x) < x$ so that the process may be continued indefinitely and the sequence of iterates satisfies

$$a < N^{(k)}(x) < N^{(k-1)}(x) < \ldots < N(x) < x.$$

Observe that

$$x - N(x) = \frac{1}{2} (x - \frac{a^2}{x})$$

for every x. Now suppose that $x-N(x) = r$ and let us compute the difference

192

$N(x) - N^{(2)}(x)$. The relation $x - N(x) = r$ is equivalent to $x^2 - 2rx - a^2 = 0$. If we set $y = N(x)$ we have $y = x - r$ and

$$y - N(y) = \frac{1}{2}\left(y - \frac{a^2}{y}\right) = \frac{1}{2y}(y^2 - a^2) = \frac{1}{2y}\left((x-r)^2 - a^2\right)$$

$$= \frac{1}{2y}(x^2 - 2rx + r^2 - a^2) = \frac{r^2}{2y} = \frac{r^2}{2(r^2 + a^2)^{1/2}}.$$

If we denote by w the function

$$r \mapsto \frac{1}{2}\frac{r^2}{(r^2 + a^2)^{1/2}}$$

defined for positive r, then what we have proved may be reformulated as follows: if $x > a$ and $x - N(x) = r$ then $N(x) - N^{(2)}(x) = w(r)$. It follows that

$$x - N^{(n)}(x) = r + w(r) + \ldots + w^{(n-1)}(r)$$

so that the problem of computing the sum $r + w(r) + \ldots + w^{(n-1)}(r)$ reduces to that of finding an explicit formula for $N^{(n)}(x)$.

We shall show later that it suffices to deduce an explicit formula for $N^{(n)}(x)$ in the case $a = 1$ only; the general case may be easily reduced to this one by a simple substitution.

Let g be the function defined for positive x by

$$g(x) = \frac{1}{2}\left(x + \frac{1}{x}\right);$$

observe that $g(x) > 1$ for $x > 1$. We begin with a lemma concerning a sequence defined recursively by means of g.

14.1 **Lemma** *Let $x_0 > 1$ and define a sequence $(x_n)_{n \geq 0}$ by the relation* $x_{n+1} = g(x_n)$. *Then*

$$x_n = \frac{(x_0 + 1)^{2^n} + (x_0 - 1)^{2^n}}{(x_0 + 1)^{2^n} - (x_0 - 1)^{2^n}}.$$

Proof Clearly it is sufficient to verify this formula inductively. We intend, however, to indicate a heuristic approach to the result. We look for solutions of the form $x_n = \frac{u_n}{v_n}$; the relation to be satisfied becomes

$$\frac{u_{n+1}}{v_{n+1}} = \frac{1}{2}\left(\frac{u_n}{v_n} + \frac{v_n}{u_n}\right) = \frac{1}{2}\frac{u_n^2 + v_n^2}{u_n v_n} = \frac{(u_n + v_n)^2 + (u_n - v_n)^2}{(u_n + v_n)^2 - (u_n - v_n)^2}.$$

Upon setting $u_n + v_n = p_n$, $u_n - v_n = q_n$, we may reformulate the relation in the following form

$$\frac{p_{n+1} + q_{n+1}}{p_{n+1} - q_{n+1}} = \frac{p_n^2 + q_n^2}{p_n^2 - q_n^2}.$$

This will be satisfied if we set $p_{n+1} = p_n^2$ and $q_{n+1} = q_n^2$. Hence

$$x_n = \frac{p^{2^n} + q^{2^n}}{p^{2^n} - q^{2^n}}$$

for suitable p and q. A possible choice is to take p and q such that $p + q = x_0$ and $p - q = 1$. This leads to the formula

$$x_n = \frac{(x_0 + 1)^{2^n} + (x_0 - 1)^{2^n}}{(x_0 + 1)^{2^n} - (x_0 - 1)^{2^n}}.$$

There is an even shorter way of obtaining the explicit formula for x_n, that based on the use of the Cayley transform. For $x \neq 1$ let

$$k(x) = \frac{x + 1}{x - 1}$$

so that $k(x) \neq 1$. The composition $k \circ k$ is the identity function. It is easy to verify that, for $x > 1$,

$$k(g(x)) = (k(x))^2.$$

If we denote by s the squaring function

$$s(x) = x^2$$

the above identity may be rewritten in the form

$$k \circ g = s \circ k$$

whence

$$g = k \circ k \circ g = k \circ s \circ k$$

194

and, for any natural number n

$$g^{(n)} = k \circ s^{(n)} \circ k.$$

Given $x_0 > 1$ write y_0 for $k(x_0)$ and observe that

$$g^{(n)}(x_0) = k(s^{(n)}(y_0)) = k(y_0^{2^n}) = \frac{y_0^{2^n} + 1}{y_0^{2^n} - 1}$$

which is nothing more than the formula obtained above. □

    Passing from the normalized case to the general case we have the following result.

14.2 Proposition *Suppose that $y_0 > a > 0$ and that the sequence y_n is defined by the recursive formula*

$$y_{n+1} = \frac{1}{2}\left(y_n + \frac{a^2}{y_n}\right).$$

Then

$$y_n = a\,\frac{(y_0 + a)^{2^n} + (y_0 - a)^{2^n}}{(y_0 + a)^{2^n} - (y_0 - a)^{2^n}}.$$

Proof Define a sequence $(x_n)_{n \geq 0}$ by the formula $x_n = a^{-1}y_n$. We have then $x_0 > 1$ and

$$x_{n+1} = \frac{1}{2}\left(a^{-1}y_n + \frac{a}{y_n}\right) = \frac{1}{2}\left(x_n + \frac{1}{x_n}\right)$$

so that

$$y_n = a\,x_n = a\,\frac{(x_0 + 1)^{2^n} + (x_0 - 1)^{2^n}}{(x_0 + 1)^{2^n} - (x_0 - 1)^{2^n}}$$

$$= a\,\frac{(y_0 + a)^{2^n} + (y_0 - a)^{2^n}}{(y_0 + a)^{2^n} - (y_0 - a)^{2^n}}.$$ □

14.3 Theorem *Let $a > 0$. Then the function w defined by*

$$x \rightarrow \frac{x^2}{2(x^2 + a^2)^{1/2}}$$

is a rate of convergence on the whole positive axis. For each natural number n and each r > 0 we have:

$$r + w(r) + \ldots + w^{(n-1)}(r) = \xi - a \frac{(\xi+a)^{2^n} + (\xi-a)^{2^n}}{(\xi+a)^{2^n} - (\xi-a)^{2^n}} = \xi - a \frac{1 + \theta^{2^n}}{1 - \theta^{2^n}}$$

$$= \xi - a \, \text{cth}(2^{n-1}\phi),$$

$$w^{(n)}(r) = \frac{2a(2r\xi)^{2^n}}{(\xi+a)^{2^{n+1}} - (\xi-a)^{2^{n+1}}} = \frac{2a\theta^{2^n}}{1 - \theta^{2^{n+1}}} = \frac{a}{\text{sh}(2^n\phi)},$$

$$s(w^{(n)}(r)) = \frac{2a(\xi-a)^{2^n}}{(\xi+a)^{2^n} - (\xi-a)^{2^n}} = \frac{2a\theta^{2^n}}{1 - \theta^{2^n}} = \frac{a \cdot \exp(-2^{n-1}\phi)}{\text{sh}(2^{n-1}\phi)},$$

where

$$\xi = r + (r^2 + a^2)^{1/2}, \quad \theta = \frac{(r^2+a^2)^{1/2} - a}{r}, \quad \phi = -\log_e \theta,$$

$$\text{sh } x = \frac{1}{2}(e^x - e^{-x}), \quad \text{cth } x = \frac{e^x + e^{-x}}{e^x - e^{-x}}.$$

<u>Proof</u> Let us denote by N(z) the Newton step at the point z for the function $f(t) = t^2 - a^2$, so that $N(z) = \frac{1}{2}(z + a^2/z)$. We have shown that

$$r + w(r) + \ldots + w^{(n-1)}(r) = x - N^{(n)}(x)$$

if x is such that x-N(x) = r and x > a. It is easy to verify that $x = \xi := r + (r^2 + a^2)^{1/2}$ satisfies both these requirements. Thus the first two assertions are immediate consequences of the preceding lemma. The third assertion can be easily obtained upon observing that the sum $\sum_{0}^{n-1} w^{(j)}(r)$ converges to $r + (r^2 + a^2)^{1/2} - a$. □

The explicit forms for $s(w^{(n)}(r))$ given in the above theorem are related to different results concerning the a priori error estimates for Newton's method. In [50] it has been shown that $|x_n - x^*| \leq s(w^{(n)}(r_0))$. The formula

$$s(w^{(n)}(r)) = \frac{2a(\xi-a)^{2^n}}{(\xi+a)^{2^n} - (\xi-a)^{2^n}}$$

can immediately be derived from the results proved in [47]. The explicit
forms of the a priori error estimates for Newton's method

$$\frac{2a\;\theta^{2^n}}{1 - \theta^{2^n}} \quad \text{and} \quad \frac{a\;\exp(-2^{n-1}\phi)}{\text{sh}(2^{n-1}\phi)}$$

appear respectively in Gragg and Tapia [15] and Ostrowski [26].

The rest of this section will be devoted to establishing explicit
expressions for the partial sums in the case of the rate of convergence

$$w(r) = r\,\frac{b + r - u}{b - r + u}$$

where $u = ((b + r)^2 - 4a)^{1/2}$ and a,b are two non-negative constants satisfying
$b^2 - 4a \geq 0$.

Again, it is obvious that we could have limited ourselves to starting the
results only since all of them may be verified without difficulty by
induction. This would, however, give no indication as to how they have been
obtained.

A lemma about continued fractions

We prove first a simple result about the function $z \to \frac{a}{b - z}$; we obtain an
explicit formula for a recursively defined sequence which will be used later.

14.4 <u>Lemma</u> *Let a and b be two positive numbers such that $b^2 > 4a$. Suppose
that z_0 is a positive number which does not belong to a countable set of
exceptional values E to be described below. Then it is possible to define a
sequence z_k such that $z_k = \dfrac{a}{b - z_{k-1}}$ for $k = 1,2,\ldots$. The sequence z_k may be
expressed explicitly as follows*

$$z_n = \lambda_2\,\frac{1 - A\,q^{n-1}}{1 - A\,q^n}$$

where

$$\lambda_1 = \frac{1}{2}\,(b + \sqrt{b^2 - 4a})$$

$$\lambda_2 = \frac{1}{2}\,(b - \sqrt{b^2 - 4a})$$

$$q = \frac{\lambda_2}{\lambda_1}, \quad A = \frac{\lambda_2 - z_0}{\lambda_1 - z_0}.$$

The exceptional set E *consists of the following numbers*

$$e_n = \lambda_1 \frac{1-q^{n+1}}{1 - q^n}$$

for n = 1,2,... . *All these numbers are different from each other and sat-isfy the inequalities*

$$\lambda_2 < \lambda_1 < e_n < b \quad for \quad n = 2,3,\dots .$$

For n = 1 *we have* $e_1 = b$.

Proof Of course, it would be sufficient to verify the formula inductively. We intend, however, to indicate one of the possible methods of obtaining the result.

Define a sequence $d_n(z_0)$ by the following formula

$$d_0(z_0) = 1$$

$$d_1(z_0) = b - z_0$$

$$d_n(z_0) = \det \begin{pmatrix} b, & -1, & 0, \ \dots & & & 0 \\ -a, & b, & -1, & 0, \ \dots & & 0 \\ 0, & -a, & b, & -1, & 0, \ \dots & 0 \\ \cdot & \cdot & \cdot & \cdot & \cdot & \cdot \\ 0, & 0, & 0, & \dots, & -a, & b, & -1 \\ 0, & 0, & 0, & \dots, & 0, & -a, & b-z_0 \end{pmatrix}$$

It follows from the theory of continued fractions that $z_n = a d_{n-1}(z_0)(d_n(z_0))^{-1}$. Upon expanding the determinant along the first column we obtain the following recursive formula for d_n

$$d_n(z_0) = b d_{n-1}(z_0) - a d_{n-2}(z_0).$$

The characteristic polynomial $\lambda^2 - b\lambda + a$ has two different positive roots

$$\lambda_1 = \frac{1}{2}(b + \sqrt{b^2 - 4a}), \quad \lambda_2 = \frac{1}{2}(b - \sqrt{b^2 - 4a}).$$

It follows that

198

$$d_n(z_0) = \alpha_1 \lambda_1^n + \alpha_2 \lambda_2^n$$

the coefficients α_1 and α_2 being determined from the initial conditions. We have

$$\alpha_1 = \frac{h_0 + b - 2z_0}{2h_0} , \quad \alpha_2 = \frac{h_0 - b + 2z_0}{2h_0} ,$$

where we write for brevity h_0 for $(b^2 - 4a)^{1/2}$.

Hence

$$z_n = a \frac{\alpha_1 \lambda_1^{n-1} + \alpha_2 \lambda_2^{n-1}}{\alpha_1 \lambda_1^n + \alpha_2 \lambda_2^n} = \frac{a}{\lambda_1} \frac{1 + \frac{\alpha_2}{\alpha_1} q^{n-1}}{1 + \frac{\alpha_2}{\alpha_1} q^n}$$

with $q = \dfrac{\lambda_2}{\lambda_1}$. Since $a = \lambda_1 \lambda_2$ and

$$\frac{\alpha_2}{\alpha_1} = \frac{z_0 - \frac{1}{2} b + \frac{1}{2} h_0}{-z_0 + \frac{1}{2} b + \frac{1}{2} h_0} = \frac{z_0 - \lambda_2}{-z_0 + \lambda_1} = - \frac{\lambda_2 - z_0}{\lambda_1 - z_0}$$

we obtain the desired formula.

It should be remarked here that the method based on the theory of continued fractions is far from being the only possible way of obtaining the above result. Instead of using the results of the theory of continued fractions its methods may be applied directly since they simplify considerably in our case.

We look for solutions of the form $z_n = \dfrac{u_n}{v_n}$ for suitable u_n and v_n. The relation to be fulfilled is

$$\frac{u_{n+1}}{v_{n+1}} = \frac{a}{b - \dfrac{u_n}{v_n}} = \frac{a v_n}{b v_n - u_n} ;$$

this will hold if the u_n and v_n satisfy the following relations

$$u_{n+1} = a v_n \quad v_{n+1} = b v_n - u_n.$$

It follows that it suffices to find v_n satisfying $v_{n+1} - b v_n + a v_{n-1} = 0$ and then set $u_n = a v_{n-1}$. Hence a solution may be obtained if we set

199

$$z_n = \frac{a\,v_{n-1}}{v_n}$$

where v_n is a suitable solution of the above difference equation. Its characteristic polynomial being $x^2 - bx + a$, every solution of the difference equation is of the form

$$v_n = \alpha_1\,\lambda_1^n + \alpha_2\,\lambda_2^n.$$

The rest follows in the same manner as above.

The partial sums

In this section we use the method developed in [49] to compute the partial sums of the infinite series for s.

14.5 <u>Lemma</u> *Let a and b be two positive numbers such that $b^2 > 4a$. Set*

$$z_* = \frac{1}{2}\,(b - \sqrt{b^2 - 4a}).$$

For $z < z_*$ the function $M(z) = \dfrac{a}{b-z}$ has the following properties:

1. $z < M(z) < z_*$ *for all $z < z_*$.*

2. *if $x > 0$ and if $z < z_*$ is such that $M(z) - z = x$ then*

$$M(z + x) - (z + x) = x\,\frac{b + x - \sqrt{(b + x)^2 - 4a}}{b - x + \sqrt{(b + x)^2 - 4a}}$$

<u>Proof</u> Since $M(z) - z = \dfrac{z^2 - bz + a}{b-z}$ and $z_*^2 - bz_* + a = 0$ we have

$$M(z) - z = \frac{z^2 - bz + a - (z_*^2 - bz_* + a)}{b - z} = \frac{(z - z_*)(z + z_* - b)}{b - z}$$

and $z - z_* < 0$, $z + z_* - b < 2z_* - b < 0$. Hence $M(z) - z > 0$ for $z < z_*$. Similarly,

$$z_* - M(z) = \frac{1}{b-z}\,(z_* b - z z_* - a + (z_*^2 - bz_* + a)) = z_*\,\frac{z_* - z}{b - z} > 0.$$

To prove the second part suppose that $z < z_*$ and that $M(z) - z = x > 0$.

200

Write z' for z + x = M(z) and observe that a - z'(b - z) = 0. Using this, we obtain

$$M(z') - z' = \frac{a - z'(b - z')}{b - z'} = \frac{z'x}{b - z'} = x \frac{z + x}{b - x - z} .$$

To eliminate z, we observe that a - (z + x)(b - z) = 0 or $z^2 - (b-x)z + a - bx = 0$. It follows that z must have one of the values

$$\tfrac{1}{2}(b - x \pm \sqrt{(b + x)^2 - 4a});$$

the plus sign yields a value $z > z_*$. It follows that $z = \tfrac{1}{2}(b - x - h)$ where $h = ((b + x)^2 - 4a)^{1/2}$ so that

$$M(z') - z' = x \frac{z + x}{b - x - z} = x \frac{b + x - h}{b - x + h} . \qquad \square$$

14.6 <u>Theorem</u> *Let a and b be two positive numbers such that $b^2 > 4a$. For each $x > 0$ let $h = ((b + x)^2 - 4a)^{1/2}$ and write h_0 for $(b^2 - 4a)^{1/2}$; for each $x > 0$ set*

$$w(x) = x \frac{b + x - h}{b - x + h} .$$

Then w is a rate of convergence on the whole positive axis and possesses the following properties

1. *For each natural number n and each $x > 0$ we have*

$$x + w(x) + \ldots + w^{(n)}(x) =$$

$$\tfrac{1}{2}(b - h_0) \cdot \frac{1 - \dfrac{x + h - h_0}{x + h + h_0}\left(\dfrac{b - h_0}{b + h_0}\right)^n}{1 - \dfrac{x + h - h_0}{x + h + h_0}\left(\dfrac{b - h_0}{b + h_0}\right)^{n+1}} - \tfrac{1}{2}(b - x - h)$$

2. *The infinite sum equals*

$$s(x) = \tfrac{1}{2}(x + h - h_0).$$

<u>Proof</u> Clearly the second statement is an immediate consequence of the first. Now suppose that $z_0 < z_*$ is such that $M(z_0) - z_0 = x$. According to Lemma

14.5,

$$M^{(2)}(z_0) - M(z_0) = w(x) \text{ and } M^{(k)}(z_0) - M^{(k-1)}(z_0) = w^{(k-1)}(x)$$

for all $k \geq 1$. It follows that

$$x + w(x) + \ldots + w^{(n)}(x) = M^{(n+1)}(z_0) - z_0.$$

In view of Lemma 14.4 it remains to prove the existence of z_0. It is not difficult to verify, however, that

$$z_0 = \frac{1}{2}(b - x - h)$$

satisfies all the requirements. This completes the proof. The formula obtained in Lemma 14.4 makes it possible to give also an explicit expression for $w^{(n)}(x)$. □

References

[1] G.R. Allan, A.M. Sinclair: Power factorization in Banach algebras with a bounded approximate unit, Studia Math. 56 (1976), 31-38.

[2] W.E. Bosarge, P.L. Falb: A multipoint method of third order, J. Optimiz. Theory Appl., 4 (1969), 156-166.

[3] W.E. Bosarge, P.L. Falb: Infinite dimensional multipoint methods and the solution of two point boundary value problems, Numer. Math., 14 (1970), 264-286.

[4] R.P. Brent: Some efficient algorithms for solving systems of nonlinear equations, SIAM J. Numer. Anal., 10 (1973).

[5] C. Brezinski: Comparaison de suites convergentes, Revue Francaise d'Informatique et de Recherche Operationelle, 2 (1971), 95-99.

[6] W. Burmeister: Inversionsfreie Verfahren zur Lösung nichtlinearer Operatorgleichungen, ZAMM 52 (1972), 101-110.

[7] P.J. Cohen: Factorization in group algebras, Duke Math. J. 26 (1959), 199-205.

[8] J.E. Dennis: On the Kantorovich hypothesis for Newton's method, SIAM J. Numer. Anal., 6 (1969), 493-507.

[9] J.E. Dennis: Toward a unified convergence theory for Newton-like methods, in Nonlinear Functional Analysis and Applications, L.B. Rall, Ed., Academic Press, New York, 1971.

[10] J.E. Dennis, J.J. Moré: A characterization of superlinear convergence and its application to Quasi-Newton methods, Mathematics of Computation, 28, (1974), 549-560.

[11] P. Deuflhard and G. Heindl: Affine invariant convergence theorems for Newton's method and extensions to related methods, SIAM J. Numer. Anal., 16 (1979), 1-10.

[12] J. Dixmier: Les C*-algèbres et leurs représentations, Gauthier-Villars, Paris 1969.

[13] M. Fiedler: Some estimates of the proper values of matrices. J. Soc. Ind. and Appl. Math. 13 (1965), 1-5.

[14] M. Fiedler, V. Pták: Estimates and iteration procedures for proper

values of almost decomposable operators, Czech. Math. J. 89 (1964), 593-608.

[15] W.B. Gragg, R.A. Tapia: Optimal error bounds for the Newton-Kantorowich theorem, SIAM J. Numer. Anal., 11 (1974), 10-13.

[16] L.W. Johnson, D.R. Scholz: On Steffensen's method. SIAM J. Numer. Anal. 5 (1968), 296-302.

[17] L.V. Kantorovich: Majorant principle and Newton's method, Dokl. Akad. Nauk SSSR, 76 (1951), 17-20.

[18] I. Korec: On a problem of V. Pták, Čas. pro pěst. mat. 103 (1978), 365-379.

[19] H.-J. Kornstaedt: Funktionalungleichungen und Iterationsverfahren, Aequationes Math. 13 (1975), 21-45.

[20] J. Křížková, P. Vrbová: A remark on a factorization theorem, Comm. Math. Univ. Carol. 15 (1974), 611-614.

[21] P. Laasonen: Ein überquadratisch konvergenter iterativer Algorithmus. Ann. Acad. Sci. Fenn-Ser.A I, 450 (1969), 1-10.

[22] G.J. Miel: An updated version of the Kantorovich theorem for Newton's method. Technical Summary Report, Mathematical Research Center, University of Wisconsin Madison, 1980.

[23] J. Moser: A new technique for the construction of solutions of non-linear differential equations. Proc. Nat. Ac. Sci. U.S.A., 47 (1961), 1824-1831.

[24] J. Moser: A rapidly convergent iteration method and nonlinear partial differential equations. Ann. Scuola Norm. Pisa, 20 (1966), 265-315.

[25] J.M. Ortega, W.C. Rheinboldt: Iterative solution of nonlinear equations in several variables, Academic Press, New York and London, 1970.

[26] A.M. Ostrowski: La méthodede Newton dans les espaces de Banach, C.R. Acad. Sci. Paris 272(A), (1971), 1251-1253.

[27] A.M. Ostrowski: Solution of equations in Euclidian and Banach spaces, Academic Press, New York, 1973.

[28] H. Petzeltová, P. Vrbová: An overrelaxed modification of Newton's method, Revue Roumaine des Mathématiques 22 (1977), 959-963.

[29] H. Petzeltová, P. Vrbová: A remark on small divisor problems, Czech. Math. J. 103 (1978), 1-12.

[30] F.-A. Potra: On a modified secant method, Math. Rev. Anal. Numer. Theor. Approximation, Anal. Numer. Theor. Approximation, 8 (1979),

203-214.

[31] F.-A. Potra: A characterization of the divided differences of an operator which can be represented by Riemann integrals, Math., Rev. Anal. Numer. Theor. Approximation, Anal. Numer. Theor. Approximation, 9 (1980), 251-253.

[32] F.-A. Potra: An application of the Induction method of V. Pták to the study of Regula Falsi, Aplikace Matematiky 26 (1981), 111-120.

[33] F.-A. Potra: The rate of convergence of a modified Newton's process, Aplikace matematiky 26 (1981), 13-17.

[34] F.-A. Potra: An error analysis for the secant method. Numer. Math., 38 (1982), 427-445.

[35] F.-A. Potra: On a class of iterative procedures for solving nonlinear equations in Banach spaces. Preprint Series in Mathematics, No. 67/ 1980.

[36] F.-A. Potra: On the aposteriori error estimates for Newton's method. Preprint Series in Mathematics, INCREST, Bucharest, No. 19/1981 (to appear in Beiträge Numer. Math.).

[37] F.-A. Potra: A general iterative procedure for solving nonlinear equations in Banach spaces, Preprint Series in Mathematics, INCREST, Bucharest, No. 37/1981.

[38] F.-A. Potra: On the convergence of a class of Newton-like methods. Preprint Series in Mathematics, INCREST, Bucharest No. 22/1982.

[39] F.-A. Potra, V. Pták: Nondiscrete induction and a double step secant method, Math. Scand. 46 (1980), 236-250.

[40] F.-A. Potra, V. Pták: On a class of modified Newton processes, Numer. Func. Anal. and Optimiz. 2 (1980), 107-120.

[41] F.-A. Potra, V. Pták: Sharp error bounds for Newton's process, Numer. Math. 34 (1980), 63-72.

[42] F.-A. Potra, V. Pták: A generalization of Regula Falsi, Numer. Math. 36 (1981), 333-346.

[43] V. Pták: Some metric aspects of the open mapping theorem, Math. Annalen 165 (1966), 95-104.

[44] V. Pták: A quantitative refinement of the closed graph theorem, Czech. Math. J. 99 (1974), 503-506.

[45] V. Pták: A theorem of the closed graph type, Manuscripta Math. 13 (1974), 109-130.

[46] V. Pták: Deux théorèmes de factorisation, Comptes Rendus, Acad. Sci. Paris 278 (1974), 1091-1094.

[47] V. Pták: Concerning the rate of convergence of Newton's process, Comm. Math. Univ. Carolinae 16 (1975), 699-705.

[48] V. Pták: A modification of Newton's method, Čas. pro pěst. mat. 101 (1976), 188-194.

[49] V. Pták: Nondiscrete mathematical induction and iterative existence proofs, Linear Algebra and its Applications, 13 (1976), 223-236.

[50] V. Pták: The rate of convergence of Newton's process, Numer. Math. 25 (1976), 279-285.

[51] V. Pták: Nondiscrete mathematical induction, in: General Topology and its Relations to Modern Analysis and Algebra IV., 166-178, Lecture Notes in Mathematics 609, Springer Verlag 1977.

[52] V. Pták: What should be a rate of convergence, RAIRO, Analyse Numérique 11 (1977), 279-286.

[53] V. Pták: Stability of exactness, Comm. Math. (Poznań) 21 (1978), 343-348.

[54] V. Pták: A rate of convergence, Numer. Func. Anal. and Optimiz. 1 (1979), 255-271.

[55] V. Pták: Factorization in Banach algebras, Studia Math. 65 (1979), 279-285.

[56] W. Rudin: Representation of functions by convolutions, J. Math. Mech. 7 (1958), 103-115.

[57] V.E. Samanskii: On a modification of Newton's method. (in Russian) Ukrain. Mat. J., 19 (1967), 133-138.

[58] J.W. Schmidt: Die Regula falsi für Operatoren in Banachräumen. ZAMM 41 (1961), 61-63.

[59] J.W. Schmidt: Eine Übertragung der Regula Falsi auf Gleichungen im Banachraum, I, II, Z. Angew. Math. Mech., 43 (1963), 1-8, 97-110.

[60] J.W. Schmidt: Regula-Falsi Verfahren mit konsistenter Steigung und Majoranten Prinzip, Periodica Mathematica Hungarica, 5 (1974), 187-193.

[61] J.W. Schmidt, H. Schwetlick: Ableitungsfreie Verfahren mit höherer Konvergenzgeschwindigkeit, Computing, 3 (1968), 215-226.

[62] E. Schröder: Über unendlichviele Algorithmen zur Auflösung der Gleichungen. Math. Annalen, II (1870), 317-369.

[63] J. Schröder: Nichtlineare Majoranten beim Verfahren der schrittweisen Näherung. Arch. Math., 7, (1956), 471-484.

[64] H. Schwetick: Numerishe Lösung nichtlinearer Gleichungen. VEB. Deutscher Verlag der Wissenschaften, Berlin 1979.

[65] A.S. Sergeev: On the method of chords (in Russian), Sibirsk. Mat. J., 11 (1961), 282-289.

[66] J.L. Taylor: The analytic calculus for several commuting operators. Acta Math. 125 (1970), 1-38.

[67] J.F. Traub: Iterative methods for the solution of equations, Prentice Hall, Englewood Cliffs, New Jersey, 1964.

[68] S. Ulm: Majorant principle and the method of chords (in Russian) Izv. Akad. Nauk Eston. SSR, Ser. Fiz. -Mat. 13 (1964), 217-227.

[69] S. Ulm: On the generalized divided differences I, II. Izv. Akad. Nauk Eston. SSR, Ser. Fiz. -Mat. 16 (1967), 13-26, 146-156.

[70] M.A. Wolfe: Extended iterative methods for the solution of operator equations, Numer. Math., 31 (1978), 153-174.

[71] J. Zemánek: A remark on transitivity of operator algebras Čas. pro pest. mat. 100 (1975), 176-178.